# Every Touchpoint

*How the Smallest Moments Impact the People, Places & Brands We Love*

ELIZABETH LORD

LANGWORTHY LORD PRESS

Every Touchpoint: How the Smallest Moments Impact the People, Places & Brands We Love



Published by Langworthy Lord Press · Greenville, South Carolina

www.langworthylord.com

ISBN 979-8-9973026-0-3 (paperback)

First Edition, 2026

# Praise for Elizabeth Lord

*"I saw firsthand how Elizabeth Lord transformed Birkdale Village from a former dairy farm into something far more meaningful: a town square where people could gather, connect and belong. In Every Touchpoint, she reveals the intentional thinking behind that achievement and shows how thoughtfully designed moments can turn a place into a community, a brand into an experience and an ordinary interaction into a lasting memory."*

— PHILIP TATE, APR, FELLOW PRSA, PRESIDENT, PHILIP TATE STRATEGIC COMMUNICATIONS, LLC

*"My experience with Elizabeth Lord ... was sufficient to recognize that when it comes to positioning, selling, promoting and growing a product or project, she provides the whole package. The Wharf was an expansive, multifaceted project and included a mix of condominium, retail and entertainment assets — all new build and coming out of the ground in rapid-fire succession. Elizabeth was able to blend these elements together seamlessly — from ad messaging and product integration to spectacular grand opening events to ongoing public relations and community outreach. She is*

*thoughtful, creative, energetic, thoroughly schooled in real estate, hospitality and consumer marketing, a wonderfully gifted writer."*

— BOB NELSON, MANAGING DIRECTOR AT THE WHARF

*"I worked with Elizabeth Lord when she was responsible for the daunting task of marketing Birkdale Village in its infancy. Elizabeth was an integral part of why Birkdale Village became such a community center for the Huntersville area. She planned and executed signature events to keep Birkdale as a desirable destination. As a direct result of her work, we saw thousands of people for the annual tree lightings and Fourth of July events ... I can attest to the eventing contributing to the soul of the place. Not everyone can pull off large-scale events, but Elizabeth is up to the task!"*

— JULIANNE MCCOLLUM, HEAD QUACK, YELLOW DUCK MARKETING (FORMERLY WITH CROSLAND)

*"I had the genuine pleasure of working with Elizabeth Lord at Addison Whitney, a top-tier branding agency, in 2012, and my only regret is that I didn't get to enjoy her talent and company longer! ... She's also one of the most talented consumer brand namers with whom I've worked."*

— TIFFANY JONAS, PRINCIPAL, THE AIO GROUP, LLC (FORMERLY WITH ADDISON WHITNEY)

*For my daughters,*

*This book was born from a simple truth —*
*every touchpoint shapes us.*

*Of all the experiences I have known,*
*none has shaped me more profoundly*
*than the privilege of being your mother.*

*You are my greatest joy,*
*my enduring wonder,*
*and the most glorious touchpoints of my life.*

*And to you, the reader — thank you for*
*choosing to begin this journey.*

*May these pages invite you*
*to pay attention to the moments*
*that are quietly shaping your story.*

# Contents

# Prologue

There is a stretch of Arizona where the dark has no edges. I know, because one night, after the sun had long gone down, I was standing in the middle of it — above Phoenix, at a rest stop with a hand-lettered sign warning travelers to watch for rattlesnakes, two old cars that had carried us this far on little more than optimism, and my daughters watching me, waiting to see what I would do next.

The oldest was old enough to be driving the second car herself. The younger ones were awake in the dark, quiet in the way children get when they can sense something is wrong. The man who had just towed us off the highway — a stranger dressed, of all things, for a rodeo — had pulled away, and I had watched his taillights shrink and vanish until there was only the enormous silence of the desert and the enormous sky above it.

I had no phone. I had no credit cards. I had a little cash from selling almost everything we owned, and several hundred miles of darkness in every direction.

I had no career to speak of then — not yet. I was a mother, married, moving my family across the country the way we always seemed to be moving, taking work whenever we landed somewhere long enough for me to find it, doing whatever it took to keep food on the table. Whatever I knew how to do, I had spent my life doing it three steps ahead: scanning the road, planning the next move, bracing for what might go wrong, because that is what you learn when you live in survival mode.

And now the planning had simply run out. There was nothing ahead of me to arrange, no next move to make, no way to think myself out of that lot. There was only the dark, and my girls, and the next hour.

I turned my head so the girls wouldn't see, and I let a few tears fall. And standing there — everything stripped away, nothing certain — I understood something I had spent my whole life waiting to be untrue: no one was coming. There was no help on its way, no one out in that dark whose turn it was to carry us. There was only me. Whatever happened next, I would be the one to make it.

I will tell you how that night ended later in these pages. The how matters less than what it began. Because that was the first time in my life I was fully awake inside it, instead of living out ahead of it — and though I could not have known it then, I would spend the decades that followed chasing that feeling. Only I would chase it for other people: taking empty ground and vacant buildings and turning them into places where strangers felt they belonged, building for them the very thing I had been missing in that desert, and would go on missing, for a long time, in myself.

I suspect you know some version of that gap — the quiet distance between how connected we have become and how seen we actually feel. It is the particular ache of this age, and the tools arriving now will only sharpen it.

This is a book about what I discovered that night. And what I spent the rest of my life learning to build on purpose. I came to call it presence — being fully here, and making another person feel truly seen — built from the small, deliberate moments I call touchpoints. How they are made, in a business and in a life, is the rest of this book.

The desert taught me the one thing I most want to hand you: that the moment we are most afraid of — the one with no plan, no map, no guarantee — may be the only place we are ever fully alive.

Let me show you.

# The Map, Drawn Backwards

*You cannot draw the map while you are still lost inside the territory.*

For most of my life, I thought I was looking for home.

I looked for it in houses and towns, in new beginnings and familiar places, in the feeling that sometimes arrived unexpectedly when I walked into a room and knew, without being able to explain why, that I wanted to stay.

I did not know that I was looking for something much harder to find.

I was looking for a way to feel fully present in my own life.

I would not have known to call it that then. I only knew that I was always moving toward something. Another job. Another house. Another project. Another place that might finally feel like the place. I learned to anticipate what might go wrong and prepare for it. Moving taught me to adapt. Motherhood taught me to protect. Work taught me to perform. Somewhere along the way, planning became the way I made myself feel safe.

It also became the way I missed things.

That is the strange paradox I can see now: I became very good at creating experiences that made other people feel present while remaining remarkably good at living ahead of my own.

I spent more than twenty-five years building places and brands and experiences — taking empty ground and vacant buildings and turning them into places where people wanted to gather. I learned how a bell, a light, a smell, a doorway, a familiar ritual, a handwritten note, or the person who remembers your name can change the way a place feels. People rarely remember everything that happened. They remember how a place made them feel.

And I became fascinated by why.

Why does one place stay with us for decades while another disappears almost as soon as we leave it?

Why do certain people make us feel seen with almost no effort, while others can speak to us for an hour and never quite reach us?

Why does a small gesture sometimes mean more than an expensive one?

Why do some brands become part of our stories while others remain nothing more than transactions?

Why do we return to certain places, rituals, songs, restaurants, streets, and seasons not because they are objectively better, but because they return us to a version of ourselves we want to feel again?

I spent my career learning how to answer those questions for other people.

I just didn't realize I was answering them for myself.

The places I loved most had something in common. They gave time somewhere to land. They slowed us down enough to notice one another. They made us feel we belonged to something larger than the moment we had entered. And when I understood how those feelings were made, I found they were rarely the result of one grand gesture.

They were built from small things.

A sound. A scent. A familiar face. A surprising detail. A ritual repeated often enough to become tradition. A place that remembers you. A person who notices that you are there.

Touchpoints.

Small, deliberate points of contact through which meaning accumulates until an ordinary experience becomes something we carry with us.

That realization changed the way I understood my work. But it took much longer to change the way I understood my life.

Because there is a difference between designing an experience and being present for one.

I could spend months thinking about how a stranger would feel walking into a place I had helped create — what they might see, hear, smell, notice, remember. I could build the conditions for belonging.

But I was still learning how to stop long enough to experience my own.

That is the part I could not see while I was living it.

I was building for other people the very thing I was still trying to find for myself.

The desert began to change that.

That night, with no plan left to make and no way to think myself out of what was happening, something I had spent a lifetime resisting became unavoidable — the present moment.

There was no next move.

There was only the dark, my daughters, and what was happening right then.

For the first time, I understood that presence is not the absence of fear or uncertainty. It may be what becomes possible when we stop trying to outrun them.

I would spend years trying to understand what that night had revealed.

I would find pieces of the answer in the places that raised me, in the people who knew me before I knew myself, in the objects and rituals that carried whole chapters of my life inside them. I would find it in the experiences I created for strangers — and in the moments when I finally felt those same principles from the other side, as a guest, a mother, a friend, a person standing still long enough to notice what was happening.

And eventually I began to see something else.

The search for presence is not only personal.

It is also the work of every business, brand, community, and person who wants to create an experience that matters.

Because whether we are designing a hotel lobby, a neighborhood, a retail experience, a brand, a dinner with friends, or an afternoon with someone we love, we are creating touchpoints. We are deciding what people will notice, what they will remember, and whether they leave feeling merely processed — or that they were actually there.

We live in an age of extraordinary convenience and extraordinary distraction. Technology can anticipate what we want before we know we want it. Algorithms can predict what will hold our attention. Artificial intelligence can imitate conversation, creativity, even something that looks remarkably like care.

And yet many of us have never felt more distracted, more hurried, or more unseen.

We are recognized everywhere and known almost nowhere.

Our devices remember what we click. Our brands remember what we buy. Our platforms remember where we go. Machines are becoming remarkably good at knowing what we are likely to want next.

The question I keep coming back to is much simpler:

Who is paying attention to the person we are while all of that is happening?

And perhaps the harder question is:

Are we?

That is the question beneath this book.

Can we learn to be present in the lives we are so busy designing?

Can businesses learn to create experiences that do more than move people through a transaction — experiences that make people feel something, remember something, belong to something?

Can we build places that give people more than convenience, brands that offer more than recognition, and touchpoints that tell someone, in a hundred quiet ways, you matter here?

And can the same principles that help us create those experiences for others teach us something about how to live our own lives more fully?

I believe they can.

Presence can be designed — not manufactured, not forced, and certainly not automated, but invited. It can be created through attention, intention, memory, ritual, surprise, beauty, belonging, and the smallest gestures that tell someone they have been noticed.

This is what I have spent my career learning how to build.

It is also what I have spent my life learning how to receive.

That is why this book moves backward before it moves forward.

It begins with the places and people that shaped me before I understood their significance, follows the experiences that taught me how memory works and what it means to be truly seen, and traces those lessons into a career spent building experiences for others — and into the realization that the work mattered for reasons far deeper than marketing.

Then it asks what happens when the world changes around us.

When attention becomes a commodity.

When convenience begins to replace connection.

When technology becomes capable of anticipating our every desire.

When the very things that make an experience human become the things machines are learning to imitate.

And finally, it asks what we do about it.

Because I don't believe the answer is to reject technology or retreat from the modern world. Nor do I believe we can solve the problem simply by slowing down.

I think the answer is more deliberate than that.

We have to decide what deserves our attention.

We have to decide what we want people to remember.

We have to decide which parts of our businesses should remain unmistakably human.

And we have to decide what kind of lives we want to be present for while we are still living them.

Every touchpoint shapes experience.

Experience shapes memory.

Memory shapes identity.

And identity shapes the lives — and the places, brands, businesses, and communities — we build next.

That is the map I have been drawing backwards.

I did not know I was drawing it while I was living it.

I know now.

And I want to show you what I found.

PART ONE

# The Hunger for Wonder

CHAPTER 1

# We Are Built for Wonder

*Before we understand the world, we feel it. We are built to be moved — and hungrier for it than we admit.*

WE WANT TO BE MOVED.

Beneath the errands and the deadlines and the endless scroll, there is a quieter want that never fully goes away. We want to be stopped in our tracks. We want something to reach us so completely that we forget, for a moment, to be anywhere else. We want to be wowed — engulfed by something larger than ourselves, something we did not manufacture and cannot quite explain.

And here is the strange thing about this moment we are living in. We have never had more ways to be stimulated, and we have rarely felt less moved.

We carry in our pockets a window onto every wonder of the world — every cathedral, every coastline, every aurora, every performance ever recorded. We can summon a symphony or a supernova with a thumb. And yet most of us move through our days with a low-grade hunger we struggle to name, a sense that we are consuming a great deal and being nourished by almost none of it. The supply of marvels has become infinite. Our capacity to feel them has not. If anything, it is thinning.

That gap — between how much we are shown and how little we are moved — is what this book is about. And I did not come to it as a theorist. I came to it as someone who spent years building wonder for other people, often at the very times I could least find it for myself.

I have stood on a stage in front of ten thousand people, with jets crossing overhead and an orchestra swelling behind me, having built the entire spectacle myself. And I have stood in a grocery line, in a different season of the same life, without quite enough to cover the food in my arms. I mention both now, at the start, because this book lives in the distance between them — between the wonder we can engineer for others and the presence we so often cannot find in our own lives. I have spent a career on the first. I am still, at this age, learning the second.

So when I write about the hunger for wonder, understand that I am writing from inside it, not above it.

We begin life on the other side of that gap. Before we learn to categorize, to judge, to doubt, or to protect ourselves, we are open. We receive the world with our whole selves. A shadow on the wall can feel like a discovery. A song can feel like an event. A room, a smell, a face, a story — all of it arrives with a kind of aliveness we do not yet know how to defend against.

I want to be careful here, because it would be easy to mistake this for nostalgia — for a sigh about how the world felt more magical when we were young. That is not the claim. The point is not that wonder belongs to childhood. The point is that childhood reveals what we are built for. The openness was never a phase. It was evidence. It told us, before we had language for it, that human beings are made to be reached.

Which is why the hunger never leaves us. It only goes underground.

As children, our curiosity points outward — why is the sky blue, where do the birds go, what is over the next hill. In adolescence it turns inward, and the questions become who am I, where do I belong, how do others see me. Then adulthood arrives, with its careers and mortgages and aging parents and the sheer logistical weight of a built life, and our attention narrows to what must be managed. We learn to be efficient. We begin measuring ourselves by what we accomplish rather than by what we notice. The days fill, and wonder quietly yields ground — not because we stopped needing it, but because we stopped making room for it.

And then, somewhere in the middle of a built life, most of us feel it return as an ache. We find ourselves reaching again for whatever wakes us up — travel, music, beauty, ritual, the rare moment that makes us breathe a little deeper. We feel it watching a child meet something for the first time. We feel it beside a fire, or under a sky so full of stars that the day's urgencies briefly lose their grip. The hunger we had as children never died. It was waiting for us to make room for it again.

Now set that hunger against the particular moment we are entering, because the timing is not incidental. We are about to be handed two things our ancestors could scarcely imagine: nearly unlimited knowledge, and a great deal more time. Both are on the verge of becoming abundant in ways that will reshape ordinary life. And here is the uncomfortable part, the part worth pausing on before we celebrate. We have been handed gifts like these before. We did not use them well.

A century and a half ago, the people who industrialized first worked themselves to exhaustion. In 1870, a typical worker in those countries put in more than three thousand hours a year — the equivalent of sixty or seventy hours a week, fifty weeks a year. Over the following century, machines gave roughly half of those hours back. In Germany, annual working time fell by nearly sixty percent. In 1930, the economist John Maynard Keynes looked at this trend and predicted that within a few generations his grandchildren would work perhaps fifteen hours a week, and that humanity's great new problem would be what to do with all the leisure.

The productivity arrived exactly as he forecast. The leisure did not. The decline in working hours slowed, then stalled, then in places reversed. We did not collect the freedom we were offered. We converted it back into work, and into the buying of things, and later into the checking of screens. Offered the most extraordinary gift of time in human history, we largely declined it.

I do not say this to scold, and there is an honest reply worth hearing: perhaps we did not squander the time at all. Perhaps we simply wanted the comforts more than the hours, and chose accordingly, and that was our right. Maybe. But if that is true, it is worth asking whether we chose well — and whether we would choose the same way again, now that we have begun to see how the story turns out. For this was only the first of these gifts. A second would follow — and, as we will see, we declined that one too. And now the gift is about to be offered a third time, larger than before, and nothing about our track record suggests we are ready to receive it.

What I can offer is not a forecast. I am not a technologist, and the future does not need another confident prediction. What I can offer is what I have spent a life learning, often the hard way: how meaningful experience is actually made. Not by accident, and not by luck, but by design — through the small, deliberate points of contact through which a place, or a person, or a brand tells us whether we matter. I have built those moments for other people and watched what happens when they land. That is the lens of this book. I call those points of contact touchpoints, and I believe they are how human beings have always turned passing time into a life that can actually be felt.

And this is the hopeful part, the reason the book proceeds as an invitation rather than a warning. The capacity for wonder is not gone. It was never lost. It is native to us — we saw that in the child — and it can be recovered by anyone willing to choose it. What we are losing is not the ability to be moved. It is the conditions that allow being moved to happen: stillness, attention, time permitted to deepen instead of merely passing. Those conditions can be rebuilt. They can be designed back into a life, a home, a gathering, a business. That choice is still ours.

So this book begins where wonder begins. Because if we can understand how it is formed, how it fades, and why it keeps calling us back, we begin to understand something essential about ourselves. We are not only consumers of experience. We are made by it. And what makes us is not only what happened to us, but what touched us.

*We are built to be moved, and we are hungrier for it than we admit. Yet the very world that floods us with stimulation has made genuine wonder harder to reach. So before we can talk about how to build meaningful experience — for ourselves, for the people we love, for the people we serve — we have to understand what is competing for our attention, and why the competition has grown so fierce. The question waiting for all of us is the same one I keep having to ask myself: in a world engineered to keep us elsewhere, how do we learn to be here?*

CHAPTER 2

# The Age of Too Much

*When everything competes for our attention, the moments that matter most are often the ones we almost miss.*

THERE WAS A TIME WHEN EXPERIENCE CAME TO US IN smaller doses. We moved through the world at a slower rhythm, and a moment had room to settle before the next one arrived. We could stand still long enough to notice the texture of a room, the sound of a voice, the smell of the air after rain, the way light moved across a kitchen table in late afternoon.

Now most of life arrives all at once. We are surrounded by images, alerts, opinions, and endless streams of content competing for the same scarce resource: us. The pace is relentless. Even when nothing is happening, something is always happening.

So we adapt. We get faster. We skim. We scroll. We learn to live with partial attention, because that is what the environment seems to demand. But something is lost in the adaptation. When we divide our attention too widely, we stop receiving the full weight of any single thing. We hear the music but not the room. We see the table but not the gathering. We notice the food but miss the feeling. We move through our lives without quite taking them in.

This is the second of the great gifts I described in the last chapter, and the second time we have struggled to use one well. The first gift was time — the hours the machines handed back, which we mostly declined. The second is access. In a single generation, the internet collapsed the distances and the waiting that used to structure ordinary days. A question that once meant a trip to the library is answered before the thought finishes forming. A song, a film, a conversation with someone across the world — all instant, all frictionless, all free. By almost any measure this is a marvel. And we have spent a remarkable share of it on distraction.

I want to be precise about what went wrong, because it changes everything that follows. The problem is not simply that there is more noise. The problem is that much of the noise is engineered. The systems that fill our days were built — by some of the most sophisticated behavioral design in history — to capture exactly the attention we are trying to protect. They are optimized to keep us moving, reacting, checking. They fragment the one condition that lets experience become meaningful: our ability to stay with something long enough to feel it.

I say this without judgment of any of us, and I include myself fully, because the distinction matters. We did not lose our attention through weakness of character. To a great degree, it was taken from us, by environments designed to take it. That is not an excuse.

It is a diagnosis — and a strangely hopeful one. Because if attention can be engineered away from us, it can be engineered back. I have spent my career on the other side of that equation, building places meant to give attention back rather than steal it, and I can tell you the same craft works in both directions. The only question is which way we point it.

The cost of pointing it the wrong way is quiet, and it is real. A conversation can pass through us without being absorbed. A meal can be eaten without being remembered. A beautiful room can be entered without being felt. A moment that should have mattered can vanish before we noticed it was there.

When that happens often enough, life begins to feel thinner than it should. Not necessarily worse. Just less textured, less rooted, less complete. We may not register the loss at first, because the input never stops. But input is not experience. Information is not meaning. Stimulation is not awe.

This is the fear I carry, and I will name it plainly, because I suspect you carry some version of it too. It is not that we will forget our lives. It is that we will be too scattered to have fully lived them in the first place — that the moments will pass without ever being encoded, because we were never quite present for them. A life can technically happen and still leave almost no trace inside the person who lived it. That is the real danger of an age of too much. Not pain. Erosion. The slow disappearance of a life into the very abundance that was meant to enrich it.

The antidote is presence, and presence requires space. It cannot exist in a rushed mind. It needs enough openness to be surprised, enough stillness to be reached. Which is why so many people are quietly hungry for places, brands, and moments that slow them down. They are not only looking for convenience. They are looking for coherence — the sense that the parts of an experience belong together, that someone cared how it would feel to be there.

You know this feeling, even if you have never named it. Think of the last time you were completely where you were — not half-here and half-somewhere-else, but fully inside the moment you were living. Maybe it was a meal that made you lose track of the hour, or a conversation so absorbing you forgot the phone in your pocket, or an ordinary afternoon that somehow arrived at full volume. You were not managing it or documenting it or racing ahead of it. You were simply in it, and it reached you, and afterward it stayed.

That is presence. It is not a mystical state reserved for the disciplined few. It is the most natural thing in the world — the way we were built to meet our own lives — and we have simply been pulled out of it so relentlessly, for so long, that it has come to feel like a luxury rather than a birthright. Hold onto whatever you just remembered. Learning to make more of it, on purpose, is what the rest of this book is for.

You can watch companies stumble onto this truth. The ones people love are rarely winning on features alone. Consider something as small as opening a new Apple product. The company spends real design effort on the box itself — the resistance of the lid, the way the contents are revealed in sequence, the quiet around the object before you ever turn it on. None of that improves the device. All of it shapes how the moment feels. Apple understood that the unboxing is not packaging. It is the first sentence of a relationship. Walk into a Trader Joe's and the same instinct is at work in another key — hand-lettered signs, crew members who actually talk to you, a deliberate smallness that reads as human rather than industrial.

People do not return for the lowest price. They return because the place feels made by people who thought about how it would feel to be in it.

That is the whole game, and it is older than commerce. People forget the details of what was said, but they remember the atmosphere. They remember whether a place felt easy or tense, generous or indifferent, warm or cold. They remember whether the experience matched the promise. And when it does, trust forms — not because of a message, but because the experience itself delivered the message.

This is where the idea at the center of this book earns its name. A touchpoint is never just a detail. It is a signal. It tells us what a person, a place, or a brand believes — whether we matter, whether care was taken, whether the experience was designed with us in mind. A single well-made touchpoint can quiet the noise around it. It can create a small pocket of coherence in a chaotic world. It can make a person stop, even briefly, and feel the difference between being addressed and being reached.

And touchpoints reach far beyond business. They shape the atmosphere of our homes, the tone of our relationships, the quality of our gatherings, the rituals that give our days a shape we can return to. Wherever people come together, touchpoints are quietly telling a story about what matters.

The good news — the thread that runs through everything ahead — is that presence can be recovered. It can be practiced. It can be invited back. We begin by letting a moment finish before reaching for the next one. By choosing environments that support our senses instead of hijacking them. By building small rituals that give our lives a rhythm we can find our way back to. That is the real challenge of the age of too much. Not to escape it, but to learn how to stay human inside it.

*We end this chapter on the work of recovering our attention — and that work raises a question. What are we recovering it for? Now and then, even in the middle of our busiest, most distracted seasons, something still breaks through. A moment reaches us so completely that everything else goes quiet, and we remember, with a kind of relief, that we are still capable of being moved.*

CHAPTER 3

# The Architecture of Awe

*Awe is what quiets us when the world exceeds our expectations. The places we build can summon it — or crowd it out.*

AWE IS ONE OF THOSE THINGS WE SEEM TO KNOW instinctively, even before we can name it. It is the feeling that comes over us when something exceeds our expectations just enough to quiet us. It may arrive through beauty, scale, sound, timing, kindness, or surprise. It can happen in grand places or small ones — a mountain range, a concert hall, a child's face, a room lit just right. And when it arrives, something shifts. We become more attentive. More open. For a brief moment the noise recedes and the distractions that usually own us lose their grip. We stop moving mechanically through the day and become fully present inside it.

That is why awe matters. Not because it entertains us. Because it awakens us. And like wonder, it comes easily to us as children and then, somewhere along the way, begins to feel optional. We learn how things work, and in the learning we sometimes lose our willingness to be surprised by them. We tell ourselves we are too busy, too practical, too experienced to be impressed. And yet we keep returning to it — traveling to find it, seeking it in music and nature and architecture and ritual, building moments of wonder for our children because we want them to feel what we once felt. The longing itself is the tell. Awe is not a luxury. It is a human need, and one we have lately allowed to go unmet.

I learned how completely awe can ambush a person on a tired afternoon in Barcelona, in the spring of 2022, as the world was just emerging from the long isolation of the pandemic. My brother and I had been traveling for weeks — through Portugal, France, and Spain — and by the time we reached Barcelona I was worn down in the particular way long travel creates, overstimulated and quietly certain I had absorbed all the beauty I could hold.

My real interest there was the work of Antoni Gaudí, especially Park Güell, where architecture and landscape and mosaic blend into one of the earliest and most imaginative examples of placemaking I had ever seen. But by our final day, I was exhausted. We were wandering the crowded streets shopping for last-minute souvenirs when we found ourselves before La Sagrada Família. Tourists lined the sidewalks. Selfie sticks filled the air. Vendors shouted over one another. It felt more like a crowded attraction than a sacred place.

My brother announced he was buying tickets to go inside. I resisted. After weeks of travel and years of cathedrals, I was certain I had seen enough stained glass and stone to last a lifetime. He insisted, so we bought tickets and stood in line, and for nearly an hour I complained quietly — irritated by the crowds, the heat, my own fatigue. By the time we reached the entrance, I had already decided the experience would disappoint me.

Then we stepped inside. And everything changed.

The moment I entered the central nave, my breath caught. I was no longer standing inside a building. I was standing inside a forest of stone and light. Massive pillars rose and branched as they climbed toward a ceiling that seemed less constructed than grown. Sunlight poured through towering windows in waves of blue, gold, green, and crimson, moving slowly across the stone like light through leaves. Cool air settled around us. Voices softened on their own. The scale of the room dissolved the noise I had carried in with me. Looking up, I felt the absurd, involuntary urge to lie down on the cool floor and simply watch the light move, as if I were staring up into a canopy of trees.

This was not architecture. It was emotional transport. And what moved me most, standing there, was the realization that I had very nearly missed it. Fatigue had almost closed me off. Familiarity had almost blinded me. I had told myself I had seen it all, when the truth was that I had simply stopped expecting to be moved.

That, I have come to think, is one of the quiet dangers of adulthood, and I do not believe it is mine alone. We grow so practiced at managing our lives that we stop letting them surprise us. We arrive at the door of wonder and decide, in advance, not to go in. We are tired. We have seen something like it before. We have somewhere else to be, even when we don't. How many doors have we walked past, certain there was nothing inside worth the stop? I nearly walked past one of the most moving experiences of my life because I had decided, in advance, that I was done being amazed.

I have thought about that afternoon for years, and not only because the basilica was beautiful. I keep returning to it because it taught me how awe is actually made — and that it is made, deliberately, by someone who cared.

The best design rarely calls attention to itself. It does not make us feel managed. It makes us feel transported. And awe is almost never the product of a single dramatic detail. It emerges when sensory moments layer together so completely that we stop observing an experience and begin to feel part of it. It lives in atmosphere, texture, sound, anticipation, contrast — in the accumulation of carefully considered touchpoints that together create something larger than any one of them alone. Gaudí did not merely design a church. He designed awe. He drafted it, the way a composer drafts a swell.

This is the secret hiding inside every experience that has ever truly moved you: awe is drafted, not stumbled into. The cathedral that stops your breath, the room that makes you exhale the moment you step into it, the arrival that leaves you feeling expected rather than processed — none of it is an accident. Someone laid the sightline, tuned the light, slowed the sequence, chose the one detail that tips a moment from pleasant into unforgettable. Gaudí did it with stone and colored glass over a century of patience. But the same craft works at every scale, in far humbler materials, and I would spend my whole working life practicing it — learning to draft aliveness into ordinary places for people who would never know it had been drafted at all. The aim is never extravagance. The aim is aliveness.

Because the experiences we remember most are rarely transactional. They are transformational. For a brief moment they pull us out of ourselves and connect us to something larger —

beauty, memory, humanity, each other. And here is what I most want you to take from that crowded afternoon in Barcelona: awe does not require excess. It requires care, and it requires that we show up open enough to receive it. It can live in a cathedral, and it can live in the way a room is arranged, a meal is served, an arrival is staged so that someone feels expected rather than processed. We can build it for others. And we can refuse, however tired we are, to decide in advance that we are done being moved.

*Awe reminds us that we are still deeply responsive to beauty, atmosphere, and emotional presence. But standing before something beautiful is not the only way we are reached. Some of the touchpoints that stay with us longest do not take our breath away. They do something quieter and just as powerful: they remind us that we matter. And to understand where that capacity to read a room and notice a person comes from — at least where mine came from — we have to go back to the town where I first learned to see.*

PART TWO

# The Places and People That Make Us

CHAPTER 4

# A Town Called Camelot

*Some places teach us how to see. We spend the rest of our lives looking for what they first showed us.*

GROWING UP IN CONCORD, MASSACHUSETTS FELT A little like growing up in Camelot. It was a town shaped by history, beauty, tradition, and ideas, and for me it became the place where childhood first took root. My family had moved often — they would go on moving for most of my life — but Concord was where the story of who I was began to settle into focus.

My father was from Boston, a Harvard graduate as his father had been before him, and in a sense we had returned him to the region that formed him. He grew up in the city but spent summers outside Gloucester with an aunt while my grandparents traveled as entertainers to the old summer hotels in New Hampshire, where my grandfather played violin and my grandmother sang. Even before I understood it, I was being raised inside a family culture where music, hospitality, performance, storytelling, and ideas all lived comfortably beside one another. I did not know yet that this was unusual. I thought it was simply what a home was.

We lived in an old Victorian house with dirt sidewalks that led down the street and around the corner into downtown, a quaint New England town of beautiful old homes and a quiet sense of permanence. Our house backed up to a playground and a large park, and the schools were close enough to walk to. Childhood there was shaped by movement and freedom — walking to a friend's house, riding a bike across town, wandering downtown alone, moving through the neighborhood with the easy confidence that there were people along the way who knew you and would look out for you.

The seasons gave the town its rhythm. Winter brought ice skating at the playground and ski lessons on the hill behind the high school. Fall meant football, the whole town drawn together on Saturday mornings to cheer the local heroes. Spring filled the fields with track meets and field hockey. Summer brought morning camp and buses to Walden Pond to learn to swim. Looking back, I realize how much of life was being taught through repetition. I was learning, without knowing it, that identity is not formed by one defining moment but by patterns — by the places we return to, the rituals we inherit, and the atmosphere that quietly surrounds us.

Concord was also a town of ideas, where the Alcott house and the Old North Bridge made intellectual curiosity and the Revolution feel like part of the air itself. The stories embedded in the town carried a kind of inheritance — one I remember most from a birthday scavenger hunt at the bridge, where I first heard of the woman who watched the battle from her window and scratched what she saw into the glass with the point of her diamond ring. Whether every detail was exact or softened by folklore has never mattered to me. The stories made history feel intimate, and they taught me that legacy, courage, and freedom are not abstract virtues. They are lived values, woven into place and ritual and memory.

Our front porch became the emotional center of the house. It wrapped around the front of the home and served as my mother's gathering place, her studio, and in many ways her stage. She loved company and conversation. In those days she believed it improper for a married woman to entertain a gentleman inside the house, so the porch became the accepted place for connection — and then became far more than proper. It was where people gathered for meals and stories and long talk, where the life of the community flowed naturally into our home.

That porch taught me something I would not have words for until much later: the feeling of a place matters as much as its appearance.

My mother's hospitality was not simply kindness. It carried a distinct emotional tone. People came because they knew they would be welcomed, fed, listened to, included. The experience was consistent — predictable in the best possible way. It created emotional trust. That, I understand now, is a kind of brand, even though no one in that house would ever have used the word. There were always people moving through — neighbors, relatives, friends, visitors, sometimes complete strangers. If someone needed a meal, we served them. If they needed care, we offered it. That kind of belonging leaves an imprint on a child. It teaches you that connection is not abstract. It is practiced, through small repeated gestures that quietly tell a person: you belong here.

I want to pause on the stranger at the table, because that detail turns out to matter more than I knew. The instinct that there should always be room for one more — that a home is measured by who it can make space for — is the same instinct I would spend a career trying to build into places that had no porch and no mother at the center of them. Long before I had the language of brand or experience, I had watched it done, every day, by a woman who simply believed people should feel held. I was learning the trade without knowing it was a trade.

When we later moved to Gloucester, I was sixteen and heartbroken to leave Concord. Gloucester was beautiful in an entirely different way — dramatic and rocky, lifted above the ocean and the golf course from the highest point on the island. It never felt to me the way Concord had. But it became another chapter of home, and once again our house became a place of gathering. My father carried a family ritual into our lives there: Saturday nights meant baked beans and hot dogs, and people knew there would be food, conversation, laughter, and room to argue about ideas or religion or politics or life itself. Those evenings might have seemed ordinary at the time. They shaped me deeply. They taught me that hospitality is not really about food. It is about creating an atmosphere where people feel free to think, speak, question, and belong.

In the summers our world expanded outward — to Chautauqua, to Martha's Vineyard, to the Charles River, where my grandfather sometimes guest-conducted the Boston Pops. One evening a cousin brought a telescope, and my father, my brother, and I lay beneath the stars for hours. That night gave me a sense of scale — a reminder that there was something larger than the immediate moment, larger than the self, larger than whatever had seemed so urgent that day. For a child, that can feel almost sacred. It is, I think, the first cousin of awe.

What I did not understand then, but know now, is that all of these experiences were layering themselves quietly over time. The porch. The neighborhood. The stories. The meals. The conversations. The stars. None of them alone defined me. Together, they formed the emotional framework through which I came to understand beauty, fairness, freedom, hardship, curiosity, courage, hospitality, and belonging. They shaped how I read a room, how I respond to atmosphere, how I notice the small things, how I recognize what is authentic and what is only performed.

That is why I have always thought of myself, at heart, as a storyteller of place. Long before I had language for it, I was already living inside the understanding that every detail matters. That atmosphere matters. That ritual matters. That people remember how something feels as much as what it says. Concord gave me that way of seeing. It taught me to notice the layers, to feel the emotional power of a place, to know the difference between a moment and a meaningful experience.

And here is the part I can only see from this distance. The very sensitivity that town gave me — the ability to read a room, to feel whether a place is holding people or merely housing them — is the thing that would later let me build belonging for strangers, in developments and resorts and town squares that started as empty ground. It is also, I have come to realize, the thing that made the hard years so hard, because a person who feels atmosphere that acutely feels its absence just as sharply. The gift and the cost came from the same place. They usually do. We are shaped, all of us, by the places that raised us — and we carry that shaping into every room we enter and every experience we build for someone else, long after we have left the porch behind.

*Concord taught me to see. But seeing doesn't equate to being protected, and the storyteller's eye does not exempt anyone from the harder seasons of a life. A few years and several moves later, I would find myself in a small town in Maine, standing in a grocery line on Christmas Eve, learning a different lesson entirely — one that would take me most of my life to understand.*

CHAPTER 5

# Three Pairs of Skates

*When we cannot give those we love ease, we can still give them wonder. Sometimes that is the truer gift.*

ONE OF THE MANY MOVES OF MY LIFE TOOK US TO BETHEL, Maine, the winter my daughters were in their tweens and teens. I had taken a job at a ski resort nearby, in hospitality, and the pay was low enough that making rent each month was its own small act of faith. But I loved that job, and I have thought since about why. People arrived on Friday nights lit up with anticipation, coats still dusted with travel, already laughing, standing at the front edge of a vacation they had been waiting all year to take. There is a particular glow a person has at the beginning of something they have looked forward to. I have always been drawn to it. It would take me years to notice that I had spent most of my life building that feeling for other people, and almost none of it learning to feel it myself.

That Christmas was hard, and it was hard in a way I worked to keep invisible. I was straining to make it happen at all, even with food stamps. My girls had known the other kind of Christmas —

the kind at their grandparents' house, with a mountain of presents under a tree that seemed to have been decorated by people who had never once worried about money. This year would not look like that, and I knew it, and I set about doing what I always did, which was to plan my way around the shortfall.

I decided the gift would be ice skates. Three pairs, so they could go up to the community park in the evenings and skate under the lights. Skating was the great love of my own childhood. I spent whole winter evenings on the ice as a girl, alone under the stars, twirling to music only I could hear, and some of the purest happiness I have ever known happened out there in the cold with my breath clouding in front of me. I wanted to hand my daughters that exact magic. I wanted to give them the thing that had saved me.

On the afternoon before Christmas, I realized we needed a few more things for dinner. It would not be a grand meal, but I wanted it to feel like something, so as the sun went down I set out alone in the car toward town. The night was frigid, the way Maine nights are in December, the cold so complete it had gone perfectly clear. The snowbanks along the road were already waist-high. And driving those empty roads, I could see the trees glowing inside the houses, every roof and hedge blanketed in snow, every window a small lit room of someone else's Christmas.

It was so quiet, and so beautiful, that I slowed the car almost to a crawl. There were no other cars at all. Everyone was already home. And for a few minutes, moving alone through that hushed and shining town, I felt the very thing I was trying so hard to give my children — the held breath before Christmas morning, the ache of anticipation that is almost better than the thing itself. Tomorrow was Christmas. Santa was coming. For the length of that drive, with nothing required of me but to steer slowly through the beauty, I was entirely inside the moment. Even in the hardest season of my life, presence was available to me. I just had to be empty-handed enough to notice it.

Then I reached the little grocery, just as they were getting ready to close. I gathered the few things I needed and got in line. And standing there, I did the quiet arithmetic every person who has been truly short on money knows by heart, and I understood that I did not have enough — not enough cash, not enough left on the food stamps — to cover what was in my arms. So I did the thing you have watched actors do in movies, the thing that does not feel like a scene at all when it is happening to you. I began putting items back. One at a time. And I remember the precise texture of my relief that I did not know a single soul in that town who might be standing behind me, watching.

I drove home through the same beautiful night. After dinner we sat together and talked, the way you do on Christmas Eve, about Christmases past and what the girls were hoping Santa might bring. And that was when I heard, for the first time, that my oldest had her heart set on a guitar. I had no idea. I had been so deep inside the work of surviving that I had never heard it — or perhaps she had told me, and I had simply not been there to receive it, in the way teenage girls will tell you the most important thing exactly once and never again. Either way, I knew, sitting in that room, precisely what the morning held. I had bought three pairs of skates for a dream that was mine. And I had missed the one thing she had actually been asking me for.

I will not pretend the morning redeemed itself. I would love to tell you we found the magic in being together, in the snow outside the window, in the careful way each modest thing was wrapped.

That is not what happened. The girls were gracious, and grateful, and underneath the grace I could see the disappointment anyway, the particular disappointment you can only read on the faces of your own children. Three pairs of ice skates. And no skaters.

For a long time, I told myself the lesson of that Christmas was simple: I should have planned better. I was a planner, after all. I should have seen it coming, gotten ahead of it, managed the disappointment before it arrived the way I managed everything else. It took me the better part of my life to understand that I had drawn exactly the wrong lesson. Better planning was never going to rescue that morning. There was no arrangement of effort that could have gift-wrapped our circumstances into something they were not. The failure was not that I had planned too little. It was that all my planning kept me living somewhere up ahead — braced for the next blow, scanning the road — while the actual child in front of me was quietly telling me what she longed for, and I could not hear her over the noise of my own vigilance.

This is one of the most human mistakes there is, and that I am very far from the only one who has made it. We tell ourselves that if we plan hard enough, watch closely enough, stay ready enough, we can protect the people we love from disappointment, from loss, from the sharp edges of a life we cannot actually control. So we live with one foot permanently in the future.

And the cruelest part — the part I learned on the cold side of a grocery counter — is that the vigilance does not even work. It does not keep them safe. It only keeps us absent. We miss the guitar. We miss the child. We miss the very moment we were trying so hard to secure.

*The planning feels like love. It even feels like presence, because it is exhausting and it is aimed at the people we care about. But it is not the same as being with them. It is a kind of distraction with a noble face.*

*That is where this whole book begins for me — not in a boardroom or a case study, but on that drive home with the radio off, learning the thing I would spend the next twenty years slowly understanding.*

CHAPTER 6

# To Be Seen

*The deepest human need is not to be noticed, but to be known — and no system can quite counterfeit it.*

ONE OF THE DEEPEST HUMAN DESIRES IS THE DESIRE TO feel seen. Not simply noticed. Not tracked. Not measured. Not processed. Seen.

My grandfather understood this long before leadership books and customer-service trainings turned it into a technique. No matter where we went, he seemed to know everyone's name — and if he did not, he learned it. He spoke to hotel staff, musicians, and waitresses with the same curiosity and respect he showed anyone else. As a teenager I found it mildly embarrassing. As an adult I understand it completely. He was not collecting names. He was reminding people that they mattered. There is something deeply human about being addressed by name. It signals recognition, presence, individuality. It says, even briefly, I see you. I know you are here. I notice those moments everywhere now — the person in a name badge who brightens when thanked by name, the server or clerk whose whole posture eases after a small, genuine acknowledgment. In a world of hurried and transactional exchanges, even a tiny moment of recognition can restore a person's sense of dignity. Names carry more than identification; they are among the earliest touchpoints through which we come to understand ourselves and our place in the world.

To feel seen is to be recognized in a way that reaches past function. It is the sense that someone has paid attention not only to what we need, but to who we are. We learn who we are partly through the responses we receive. A child discovers confidence when someone believes in them. A student glimpses possibility when a teacher names a strength they had not noticed in themselves. A friend or mentor can change the course of a life simply by seeing something valuable before the person sees it in themselves.

So being seen does more than feel good. It tells us who we are.

Children know this instinctively — a child lights up when a parent notices the drawing taped to the refrigerator, or remembers a story told days earlier. Adults are no different. We only disguise the need more carefully. We still long for environments and relationships that make us feel understood, welcomed, and emotionally safe. And what stays with us is often surprisingly small: a tone of voice, a moment of patience, an atmosphere that quietly says you belong here. That feeling carries enormous power, because being seen is how trust begins.

Here is where our moment complicates an ancient need, and where the deepest argument of this book comes into focus. We are now surrounded by systems engineered to simulate recognition. Apps greet us by name. Algorithms learn our preferences. Platforms remember our habits and anticipate our choices. In many ways this is remarkable, and genuinely useful. But recognition is not connection — and that distinction is about to become one of the most important things a person, or a business, can understand.

A system can know our habits without knowing us. It can remember what we bought, what we watched, where we traveled, what time we wake, which songs we prefer, and which advertisements are most likely to hold us. It can map our behavior with astonishing accuracy. But pattern recognition is not empathy, and prediction is not care. A hotel app that remembers your preferred room temperature is useful. It is not the same as an employee who senses you are exhausted after a brutal day of travel and quietly slows the pace of the entire interaction. A streaming service can surface a song you will probably like. It cannot reproduce the feeling of a friend saying, I heard this and thought of you. One is predictive intelligence. The other is emotional presence.

Let me make this uncomfortably personal, because I suspect you will recognize it. There are people in my own family — people I love, people I would call close — who know less about the texture of my days than the phone in my pocket does. My apps know my taste, how often I visit a place, where I shop and bank and work. My bank has learned my rhythms so well that it anticipates when my next payment will land before I have even made it. I am known, in that sense, with a thoroughness no person in my life could match.

And none of it is knowing me. Knowing me online is not knowing me. It is surveillance mistaken for intimacy — a perfect map of my behavior with the actual person left out.

The strangeness tips into something worse when it reaches the tender places. I log in to check a medical result, and within hours my screens fill with advertisements for conditions I am fortunate not to have — my private worry harvested, packaged, and sold back to me before I have even finished being afraid. That is not being seen. That is being seen through. It is the opposite of the safety real recognition creates; it is exposure without care.

Which is exactly why a friend, a spouse, a grandchild who truly sees you matters more now than it ever has — because they offer the one thing the machine cannot. And here is the thought that stops me cold. When I am gone, my online profile will die with me, and it will leave behind nothing worth keeping. There will be no stories for my grandchildren to tell — not because I went unrecorded, I was recorded more completely than any generation in history — but because so much of me was turned toward a screen instead of toward the eyes of a child or the heart of a friend. A profile is not a life. Only the moments we spend truly seeing one another ever become a story worth passing on.

This matters more by the day, because as artificial intelligence grows more fluent, the counterfeit of being seen is becoming extraordinarily convincing. We are about to live among systems that can perform attentiveness flawlessly — that remember every preference, mirror our language, and respond with apparent warmth. They will be frictionless and tireless and, in their way, impressive. And they will still not know us. The danger is not that the simulation fails. The danger is that it succeeds well enough that we stop reaching for the real thing — that we accept being recognized and quietly forget what it felt like to be known.

That is the paradox of a personalized age, and I feel it as sharply as anyone: we are recognized constantly, and many of us still feel unseen. The experience can be smooth, efficient, even flattering, and remain emotionally hollow — like an acquaintance performing the role of a close friend, or a snack standing in for a meal. It satisfies a function without nourishing the need beneath it. Because the need was never simply to be known. The need is to matter.

None of this is an argument against the tools. Used well, technology can serve genuine connection — it can clear away the busywork so that human attention is freed for the things only human attention can do. The error is in asking it to replace what it can only imitate. Attentiveness, intuition, sacrifice, vulnerability, care — these remain stubbornly, irreplaceably human. And as life grows faster and more mediated, people are hungering for them more, not less. They long for conversations that are unhurried, hospitality that is sincere, rituals that create continuity, places that restore attention rather than compete for it.

The strongest touchpoints do exactly that, and they do it through emotional consistency rather than grand gestures. A favorite coffee shop is rarely beloved only for the coffee. A family gathering is meaningful not because every detail is perfect, but because the emotional tone is familiar and trustworthy. People return to experiences that reliably reinforce how they want to feel. That is true of brands, and it is just as true of friendships, traditions, and homes.

To be seen, in the end, is to feel safe enough to belong — not because someone studied us, and not because a system predicted us, but because another human being acknowledged us with care. That message rarely arrives in a single dramatic moment. It accumulates, through small repeated touchpoints that quietly say: you matter here. And in a world optimizing relentlessly to predict us, the people and places that take the harder path — that actually know us — may offer the rarest and most valuable touchpoint of all. The question each of us gets to answer, in our homes and our work, is which one we are practicing: the prediction, or the presence.

*We have been talking about being seen — what it is, and why no system can quite counterfeit it. But an idea is one thing, and a life is another. Before I go on, let me show you what being seen looks like when it lasts a lifetime — in the one friendship that has been the single constant of mine, and in what it taught me about the risk hidden inside letting anyone truly know us. Her name is Janice.*

CHAPTER 7

# Janice

*To be truly seen, we first have to let ourselves be known. The friend who does it for a lifetime becomes a kind of home.*

I CANNOT WRITE ABOUT BEING SEEN WITHOUT WRITING about Janice.

If there has been a single constant across my life — across fifty moves, a dozen reinventions, the glory days and the years I would rather forget — it has been her. Janice is the friend who has known me longest and known me best, and I have come to understand that those are not two separate things. They are one.

I met her when I was about eleven, at one of the first sleepovers of my life. A friend had invited me to her home in Conantum, a pocket of Concord unlike any I had seen. I had grown up in one of the town's beautiful old historic houses, all heavy moldings and history; Conantum was the opposite — low, glassy, clean-lined houses tucked into the woods, refreshingly spare and modern and, to my young eye, perfect. I remember lying on a heated floor beside a window that ran from floor to ceiling, watching snow fall on the leaves just inches from my face, mesmerized that so much beauty could be that close and that quiet.

That evening we walked down the street to meet another of my friend's friends — a girl named Janice. If the first house was serene and spare, this one was its glorious opposite: Janice was one of seven children presided over by a wildly creative mother, and the whole place hummed with a warm, overflowing chaos. Somewhere between those two homes — the still one and the loud one — sat my own, and standing in Janice's kitchen that night I understood something that would turn out to describe my entire life: I loved them all. I loved the quiet beauty and I loved the chaos and I loved the ordinary middle I had come from. And I knew, the way children know things long before they can defend them, that Janice was my kind of people. That has never once changed.

Janice became everything I quietly wanted to be. While I went off to a small women's college in New Hampshire, she went to Parsons in New York and became a genuinely gifted fashion designer — an artist, really, the kind who can make something beautiful out of anything she touches. She is also an exceptional cook and caterer and, for good measure, restores antiques. Once, sailing with my sister and her husband, Janice was down in the galley turning out hors d'oeuvres one minute and up on deck hoisting the sail with the crew the next, entirely at home in both. I have never met anyone quite like her. And everyone who meets her feels it — really, everyone loves Janice. She has a gift I have spent a lifetime admiring: she can make each person in front of her feel like the single most interesting human she has ever met.

She was the one who gave me my first Martha Stewart cookbook. At the time I had no money for the kind of entertaining those glossy pages imagined — but that was never the point, and Janice knew it was not the point. She had simply seen, long before I could see it in myself, the person I would one day become: someone who would care about how a table is set and how a room makes people feel. The gift was not really the book. The gift was being known that well.

And she kept knowing me, through everything. When I moved to Gloucester in high school, heartbroken to leave Concord, she came to visit again and again — as, in fact, she would come to visit me almost everywhere I ever moved, following the long trail of my relocations as though distance were only a minor inconvenience. She came to Chautauqua. She was maid of honor at my wedding. Years later she catered my daughters' weddings, assembled the bouquets, sewed the dresses. We have walked New England beaches in the dead of winter, camped for weeks with six children — five of them under six years old — on the sands of Assateague, rafted the rivers of Idaho, and wandered Europe together. We have cried and commiserated and laughed across all the miles and all the years. She knows my fears, my secrets, my history. She has seen me at my lowest, and she has seen me in my glory, and through all of it she has stayed exactly the same distance from me: close.

I have wondered, sometimes, what actually holds a friendship together across that many decades and that much distance. I think I finally know. When we find someone who genuinely sees us — not the version we perform, but the real one underneath — we hold fast, because that kind of seeing is the rarest thing there is. Janice has seen me in every phase of my life. And though she has often lived a whole country away, we always seemed to be moving through the same seasons at the same time, so that even the distance felt like company.

Here is something I have learned. Being truly seen is not only a gift. It is also a risk. To be seen — really seen — you first have to let yourself be known, and that means pulling back the drapes, the coverings, the decorations, and all the noise we hide behind, and letting another person look at what is actually there. That kind of intimacy can be intoxicating and frightening in the very same moment. And it demands the one thing this book keeps insisting is scarcest of all: presence. You cannot be truly seen from behind a performance. You have to actually show up, undefended, and let it happen.

I think a great many of us are quietly afraid of that. We will risk it, maybe, with a spouse, and otherwise keep the drapes drawn. It feels safer. It is also lonelier than we like to admit. What a lifetime with Janice has taught me is that the risk is worth everything — that to be known that completely by even one other person is a kind of home you can carry anywhere, through any number of moves. And the same is true, in its smaller and quieter way, of the places and even the businesses that somehow make us feel seen rather than processed. When someone, or something, pulls back the drapes and genuinely meets us, we never forget it. We build our lives around it.

I always wanted to be Janice. I said that for most of my life — she was the talented one, the effortless one, the one everybody loved. It is only now, writing this, that I can finally see what she was really doing all those years. She was seeing me so clearly, and so kindly, that she made it a little bit safer for me to become myself. Maybe that is the deepest thing one person can ever do for another. And maybe — I can almost say it now — I turned out to be a little like her, in my own way, after all.

*Janice came to visit me almost everywhere I ever moved — that was her gift, to keep finding me no matter where I landed. But a person is not the only thing that can see you, hold you, and hand you back to yourself. Some places do it too. There is one place that did exactly that for me, before I had any say in the matter, and it has kept a version of me safe across every season of my life, the way Janice has. It is where we go next.*

CHAPTER 8

# The Memory of Place

*Some places stay with us because we left part of ourselves there — and because, unlike so much else, they stayed.*

Some people spend their whole lives in one place. I spent much of mine leaving them. By the time I reached adulthood I had moved nearly fifty times, and I will come to what all that leaving taught me. But first, I need to tell you about the one place I never had to leave for good — the place that, more than any other, taught me what a meaningful life could feel like.

For me, that place is Chautauqua.

My family's roots run deep there. My grandmother, whom I never knew because she died before I was born, was connected to Chautauqua long before I arrived. She graduated from Smith College, became a composer, and won a national competition that led to a performance of her music for Eleanor Roosevelt at the White House. My mother grew up there and reminded me often that I was a fifth-generation Chautauquan. As a child, that history felt mildly interesting. As an adult, it feels like inheritance.

Some of my happiest memories live there. There were no cars inside the gates, and that alone changed everything — the rhythm was slower, softer, more deliberate. Summer days learning to sail and waterskiing on the lake. Bicycles across the bridge, frogs and turtles hunted in the ravines, ice cream on the way home. Evening concerts drifting across the grounds, and the plaza during the folk era of the 1960s, where we sat listening to guitar music about war and justice and change. Friendships that resumed each summer as naturally as if no time had passed between them. Even now I can close my eyes and walk those brick paths from memory. I know where they lead. I know the sound of piano drifting from the practice shacks, the clatter of plates from the cafeteria, the sudden rise of voices as a lecture let out in the amphitheater. In some strange way, I can still return there without moving at all. Nothing about those days was hurried. And yet they were never empty. They were full in the deepest sense of the word.

And through all of it, every fifteen minutes, the bells. The Westminster chimes ring from the bell tower across the grounds, marking the quarter hours and sounding the hour on the hour, carrying clear across the rooftops and out over the lake. No matter where you are — mid-conversation on a porch, halfway across the water under sail, walking a brick path at dusk — you know the time, because the time comes to you. There is no need for a watch, no reason to check a clock. You simply hear it. And without ever being told, every child on those grounds absorbs something the rest of the world is busy forgetting: that time can be musical, magical, a part of the very air you move through, rather than a number you anxiously consult. The bells do not interrupt the day. They hold it together. I did not know it then, but I was being taught the difference between checking time and living inside it —

a lesson it would take me most of my life, and a desert rest stop, to finally understand.

Let's pause on that, because it is the heart of this book hiding inside a childhood memory. Those summers were unhurried and completely full at the same time — and that combination is exactly the thing our age has lost and is so hungry to recover. We have been taught to equate fullness with speed and volume: more activity, more stimulation, more to show for the day. Chautauqua taught the opposite. It was rich precisely because it was slow enough to be felt.

Chautauqua was not only a summer place. I lived there through several winters as well, when the houses stood shuttered and the snowbanks rose so high they nearly swallowed the telephone poles. Life slowed to the pace of the weather. There I first encountered a particular kind of solitude — not loneliness, but stillness, the kind that teaches patience and invites reflection.

It was also where I first met loss. One autumn, a young acquaintance was killed in a car accident on his way home from a basketball game. Earlier that evening he had offered me a ride; my mother insisted I take the bus instead, and I did, and he never made it home. Later, a young man I had briefly dated was killed on a nearby road. Those experiences left a mark. For the first time I understood that life was not endless, that the people around us would not always remain, and that a place can hold joy and grief together in ways we rarely recognize until years later.

Most of all, Chautauqua gave me continuity. After a childhood of constant leaving, I did not have to rediscover myself there each summer. I could simply arrive and remember who I was inside that world. Enclosed on three sides by a fence and on the fourth by the lake, it felt both protected and open — we were free to roam, and the only real rule was to stay inside the gates. For a child, that mix of trust and structure is everything. It creates safety without confinement, freedom without fear. That, I think, is why I have always loved a front porch, which in Chautauqua was never just architecture but the space between private life and community life, where neighbors stopped to talk and the life of the place flowed up onto your steps.

When I think of Chautauqua now, I do not remember it simply because it was beautiful. I remember it because life happened there. Friendships formed there. Loss arrived there. Part of who I am took shape there. The place became inseparable from the moments it held. And I suspect that is true of the places you love most as well. We do not love them for their architecture. We love them because they witnessed us. They were present for our becoming, and so they hold a version of us we can return to when we need to remember who we are.

I did not know it then, but Chautauqua was quietly handing me a blueprint. Years later, when I set out to build places for other people — places that began as nothing but empty ground — I was not trying to recreate Chautauqua. I was trying to recreate the feeling: the feeling of arrival, of welcome, of belonging to something larger than yourself. I had been given that feeling for free, as a child, by a place I did nothing to earn. I would spend a career learning how hard it is to build on purpose.

*Chautauqua gave me belonging before I knew enough to be grateful for it, and it did so by staying — it was the one place I never had to leave for good. But most of the places in my life were the opposite. They shaped me by being left. Before I understood belonging, I had to survive a childhood of almost constant departure, and the lessons that came from all that leaving turned out to be just as formative as the ones that came from staying.*

CHAPTER 9

# The Places That Raise Us

*Every environment teaches us something about who we are, and who we might yet become.*

THE PLACES THAT RAISE US ARE NOT ALWAYS THE PLACES where we are born. Sometimes they are. But often they are the places we return to, the places we leave behind, the places that challenge us and quietly shape us while we are busy living our lives.

For a long time I believed the power of a place came from what it contained — the buildings, the beauty, the people, the objects. Over time I came to understand that what matters most is not the thing itself. It is the feeling a place creates in us, and the way that feeling shapes our days, our habits, our memories, and eventually our identity. Places do more than shelter us. They teach us what community feels like, what beauty feels like, what freedom and safety feel like. Some of those lessons come from staying. Others come from leaving.

I learned mostly from leaving. I grew up moving — a lot. By the time I reached adulthood I had attended more schools and lived in more places than I could easily count. Every move meant learning a new landscape, a new routine, a new set of unspoken social rules. Again and again I arrived somewhere unfamiliar and set about trying to figure out where I belonged.

People assume the hardest part of moving is leaving. Sometimes it is. But what I remember most is arriving. Walking into classrooms where everyone already knew one another. Entering conversations full of references I did not understand, laughing a beat late at jokes built on history I had not lived. I changed high schools as a sophomore and felt that acutely. Several of the most popular girls took me under their wing — gracious, welcoming, genuinely kind — and still I felt strangely outside the circle. Not because anyone excluded me, but because I lacked the shared context. I did not know the stories behind the jokes. I was present, but I was not yet part of the story. It took me years to understand what that was teaching me: belonging is not proximity. Belonging is shared experience. Shared memory. Shared history.

Moving taught me something else, too, and I learned it from my mother. She had a gift for framing every uprooting as an adventure. At the time I sometimes wished she would just admit what we were losing — the friends, the familiarity, the comfort of knowing exactly where I fit.

As an adult, I understand what she was doing. She was teaching me that uncertainty could become possibility if I met it with curiosity instead of fear — the same sleight of hand I would one day perform for my own daughters, turning a collapse into a road trip, a hard season into an adventure. It was a gift. It was also, I would come to see, a way of staying one step ahead of a moment instead of inside it. But that lesson was still years away.

Years later, a very different place taught me a different lesson. When my granddaughter faced a serious medical challenge, I moved into a small converted shed behind my daughter's home in Austin, Texas. What began as a practical arrangement became one of the most meaningful chapters of my life. The space was tiny. The life inside it was vast. Three generations worked together to hold one another up. We shared responsibilities, solved problems, adjusted schedules, and slowly found our rhythm.

And there I understood something I had somehow missed across fifty moves and far grander homes: feeling at home has almost nothing to do with square footage. It has everything to do with purpose. In that tiny shed I felt useful, needed, valued, connected. It taught me that a meaningful life requires far less room than we imagine. What it requires is belonging.

I share that here, against everything we are told about more, because it is also the quiet rebuttal to the anxieties of our age. We are about to be handed more — more time, more access, more capability than any generation before us. And the shed is my evidence that more is not the thing. The fullest season of my life unfolded in the smallest space I ever lived in, with no abundance of anything except people who needed each other. Whatever is coming, that will remain true.

Today, after a lifetime of moving, I live in a place I chose entirely for myself. Greenville was not chosen by parents, or marriage, or children, or career. It was chosen because it felt right. For the first time, I was not adapting to someone else's story. I was deliberately writing my own.

Strip away the addresses and the square footage and the moving trucks, and every one of these places was teaching me the same thing in a different key. What a place gives us is never really the building. It is a feeling — of being held, of being needed, of belonging to something larger than ourselves — and that feeling is built the way everything in this book is built: out of small, repeated moments of human contact, until a patch of ordinary ground becomes a place we cannot leave without grief.

That is why we spend so much of adulthood quietly searching for somewhere that feels like home. We are not looking for an address. We are looking for the feeling certain places once gave us — and hoping we might learn to build it again.

*I have told you now about the places that made me — the ones that stayed, the ones I left, the tiny shed that taught me the most. But places are never the whole story. We are shaped just as deeply by people, and above all by the rare few who see something in us before we can see it in ourselves. There was one such person, and I found her down on her knees on the floor of a shopping mall, making something ordinary beautiful.*

CHAPTER 10

# The People Who Raise Us

*Now and then someone sees us before we can see ourselves — and, in doing so, hands us back a life*

WE MADE IT ACROSS THE COUNTRY. THAT IS THE PART OF the story I can tell lightly now, from this distance, though it did not feel light at the time. By 1993 we had landed back in Charlotte, and the adventure was over, and what remained was the plain arithmetic of a life that needed rebuilding. I had three daughters, no job, and almost no money. The desert was behind me — a night I will come to, in time. The dark rest stop was behind me. But the question it had left me with was still sitting in my chest every morning: from here on out, it was all up to me.

Here's what that kind of low feels like, because I suspect some of you are standing in it right now, or have, or will. It is not dramatic. It is quiet and administrative. It is the daily, grinding work of figuring out how to begin again when you have run out of plan. I had spent years living braced for disaster, and disaster had come anyway, and now I had to do the thing I was worst at — stand still in my own life long enough to take the next real step.

So I went back to a place I knew. Years before, I had done some face painting at a local mall, and I remembered the woman who had run the marketing there — Donna Ashcraft, a former boss who had also become a friend and, though I would not have used the word then, a mentor. I drove to Eastland Mall to find her, not with a plan so much as an instinct that she was someone worth standing near.

I found her on her knees.

She was down on the floor of the mall's common area, planting silk flowers one by one into a display she was building by hand, making something ordinary beautiful because that was simply how she moved through the world. Eastland was a large, aging, urban regional mall with an ice skating rink in its center — not a glamorous place, not a place anyone would have called magic. And there was Donna, making it so anyway, one flower at a time.

We talked. I told her, more or less, where I was. And she offered me a position managing customer service. It was a modest job, an entry back into working life, and on paper it was not much. But here is the thing I have never forgotten, the thing that made me say yes before she had finished asking. It was not the job. It was the way she looked at me. At the lowest point of my life, with almost nothing to show for myself, Donna looked at me and saw not what I had been reduced to but what I was still capable of. She had always been able to do that. She saw something in me that I could not, at that moment, see in myself.

I have written about what it means to be truly seen. I wrote it as an idea. But I learned it as a rescue. Donna's faith — not her job offer, her faith — is one of the reasons I am able to write any of this at all.

I think most of us can name a person like that, if we are lucky: someone who saw us before we could see ourselves, and whose belief became a kind of permission. We do not rise on our own, however much the culture likes that story. We rise because somewhere along the way, somebody looked at us with more confidence than we had earned yet, and we decided not to disappoint them. That is not weakness. That is how human beings are actually built. We become who someone first believed we could be.

What I could not have known, taking that small job to keep my family afloat, was that I had just walked into the school that would shape the rest of my working life.

Twice a year, Eastland's incredible team — led by Donna, and designed and constructed by Todd Alexander — built a holiday display that was known to be one of the finest of any mall in the region. At Christmas, and again in Spring, the common area was transformed into a stage where Santa or the Easter Bunny greeted children. And unlike most malls at the time, these were not sets ordered from a catalog and dropped into place by a vendor. They were designed and built by in-house crews — our own people — who began work in July to have the Christmas set ready. There were winter castles with horse-drawn sleighs and what seemed like millions of lights. There were Alice-in-Wonderland forests with giant teapots and flowers spilling everywhere. Each season was entirely its own. Music and snow swirled through the air. An animated bear orchestra played. School children were bussed in from surrounding towns to see it, and there was a little shop where a child could buy a present for a brother or sister at prices a piggy bank could cover.

It was, simply, magic. And working inside it, season after season, I was learning something I did not yet have a name for. There is a term — sense of place — that was just then gaining momentum in the land development world, the idea that a physical space can be made to hold feeling, identity, belonging. I did not know the term. I just knew the magic, and I knew it was made, on purpose, by people who cared enough to build it by hand. I had done a smaller version of the same thing for my own daughters — in birthday parties, in trips, in a cross-country drive dressed up as an adventure. Same passion, different scale. And somewhere in those Eastland seasons I understood what I wanted to do with the rest of my life. I wanted to make magic. I wanted to build places that made people feel the way that mall, against all odds, made a busload of school children feel.

Donna believed in me twice. The first time, she gave me a way back in. The second time, she gave me a way up — though she did it by leaving. When she moved on to another opportunity, the path opened, and I became Director of Marketing at Eastland. She had not only seen me and lifted me. She had made room for me, and then stepped aside so I could rise into it. That is the rarest kind of mentorship there is. Most people who believe in you want to keep you close. The truly generous ones believe in you enough to clear the way and let you go. I have tried, in the years since, to be that for other people. I have never been sure I matched her.

I spent the next several years there, rising, learning, building seasons of magic and watching what they did to the people who walked through them. And then the world changed, and the place did too.

After September 11, 2001, Eastland was purchased by a real estate investment trust. The economics shifted, as they do, and the in-house productions — the hand-built castles, the bear orchestra, the crews who started in July — were replaced by the same seasonal package every other mall in the country was ordering from the same vendors. The magic did not fade gradually. It was swapped out, all at once, for something efficient and hollow and indistinguishable from everywhere else.

I knew, almost immediately, that it was time to leave. And I want to be clear about why, because it is the whole point. I did not leave because I had found something better. I left because I had watched the magic die, and I could not bear to stay and manage its replacement. Something in me already understood the thing this entire book is trying to say: that the difference between a place ordered from a catalog and a place built by people who care is the difference between being processed and being seen — and that once you have felt the real thing, the counterfeit becomes unbearable. The REIT did not do anything villainous. It did the ordinary, sensible, modern thing. It just did not understand what it was throwing away. Most of the time, that is exactly how magic dies. Not with malice. With a spreadsheet.

I had heard about a new project being built north of Charlotte, and one day I drove out to see it. It was a mess of construction, and my first honest thought was that it would never work. There were malls, and there were strip centers, and this was neither. It looked like a small-town main street, except it faced inward — so driving down the road, you could not even see what stores were there. Nobody is going to shop here, I thought. They are not going to get it.

And then, almost in the same breath, I thought: unless someone tells them the story.

I contacted Peter Pappas at Pappas Properties, who at the time was building Birkdale Village as a joint venture project with Crosland. I told him about my background, and I pitched him hard on something he had not asked for — a hospitality center, a place where the vision of what they were building could be told, to the families who live in the surrounding communities and to the retailers who would fill the storefronts below. He was gracious and unmoved. Thanks, he said, but we have someone. I kept going. And somewhere in the pitch I told him I could make it snow in that Nantucket-style main street village of his — real winter, manufactured on purpose, the way we had done it at Eastland before the spreadsheet arrived. That was the moment. I watched him catch the vision. I watched him see it.

I did not know yet that the place was called Birkdale Village. I only knew I had found, after the dark and the climb back and the school I had not meant to enroll in, somewhere to build the magic again — from nothing but a cow pasture and a story.

*Donna saw me when I could not see myself, and in doing so she handed me back a life — and a calling. Everything I learned at Eastland, about how a place can hold feeling and how easily that feeling can be discarded, I was about to put to the hardest test of my career. I had told a developer I could make it snow. Now I had to turn fifty-two acres of bare ground into a place that strangers would fold into their lives — and, if I was right about everything this book believes, would still be returning to a quarter-century later.*

CHAPTER 11

# What We Built

*Belonging can be built on purpose. From bare ground, for strangers, a place can be made to hold feeling.*

PETER A. PAPPAS, FOUNDER AND CEO OF PAPPAS Properties, hired me to market a shopping center that did not yet exist.

That is not quite how he would have put it. What existed was a construction site — ten miles north of Charlotte, in the lake country around Lake Norman, where old dairy farms and cow pastures were becoming neighborhoods almost overnight. What Peter saw, standing in the dust, was a town. A real one, with apartments over storefronts and a Main Street and a fountain where children would play. *Mixed-use,* the industry called it, though almost no one in the area yet understood what that meant. Peter understood. He was a visionary — a man who could stand in a field and see the life that would one day fill it, and a leader I deeply respected. He would tell me what he wanted — "How about a water fight on Main Street?" — and then leave me to figure out how on earth to make it happen.

I almost didn't get the job. Peter already had a marketing director in mind, someone from another of his projects. What changed his mind was a pitch — and the pitch itself, I realize now, was the whole argument of this book arriving before I had the words for it.

I told him he needed a Hospitality Office.

Not a leasing trailer. A place where a person could walk in off a construction site and understand what was being built here — and, more than understand it, *feel* it. We would have a model, and diagrams, and architectural renderings. Leasing could bring prospective tenants there and stand them in front of a future that had not yet begun to happen. And there would be me, telling the story of a town that existed, for now, only as an idea and a great deal of dust.

Peter went with me. I have thought about that a great deal since. He did not choose the candidate with the safer résumé. He chose the one who understood that before you can sell a place, you have to make people feel something about it — and that the feeling has to be built, deliberately, out of nothing but a story and the will to tell it.

Then he handed me the dust and told me to make it a town.

Birkdale Village had several problems, and I may as well name them. SouthPark Mall, in Charlotte, was the obvious competitor. But so was Walmart — the competitor no one likes to admit is taking their dollars. SouthPark had the experience; Walmart had the convenience. We had neither. Yet. And the architecture itself was unfamiliar here. Apartments over shops make sense in a city. We were in a cow field. Who wants to pay premium rent to live above empty storefronts and constant construction?

Some of the advice I received was to market Birkdale Village to greater Charlotte as a new and exciting shopping *destination.* I didn't believe it would work. I had learned something at Eastland Mall, in my previous life: if you cannot get the people within a two-mile radius to shop with you, you have failed. We had the people. What we lacked was a reason for them to choose us.

So my approach had to be different. I didn't want to build a destination. I wanted to build a love story — for the people who already lived nearby.

There was fear, too. Leaders in the surrounding towns, Huntersville and Cornelius especially, worried that Birkdale Village would hollow them out. I understood it; when you bring major retail into a small place, it changes the landscape. I told them that if they would tell me what they had planned, I would make sure we never competed with it. Our impact was coming either way. I wanted it to be a gift to those towns, and as little of a wound as I could manage.

I was working between the developer and its joint-venture partner, and I did not really have a full marketing budget. What I had was an advertising agency, hired before I arrived, with a strategy built almost entirely on ad spend. Buy the awareness, the thinking went, and the people will come.

I wanted to build an experience instead.

I remember the meeting where I asked the CEO of the joint-venture partner for a thirty-thousand-dollar Christmas tree and two snow machines. If he would give me those, I said, I would find the money for everything else. He did. I am still a little astonished that he did.

That tree became the center of everything.

Most children meet Santa the same way — a red suit and white gloves under fluorescent light, at the end of a long line in a crowded mall. I wanted ours to meet him under the stars, with snow falling on the tips of their noses. Our billboard on the I-77 corridor said only two words: *Snowing Nightly.* Every fifteen minutes the machines in the arbors above the village would let go, and it would snow on Main Street in North Carolina.

But an experience, I was learning, is never one thing. It is a collection of details, and you have to layer them. So we did. Carriage rides with clip-clop hoofbeats and warm blankets and costumed drivers. Holiday music under the streetlamps. Branded banners and fresh green wreaths. A small photography shop on Main Street to take the pictures.

I still needed a house for Santa to sit in. I found a man who built log cabins and made him a deal: build me a little Santa house, and I will make you part of the show. For all of October, as stores began to open and streets were paved, passersby watched a log cabin rise on Main Street — the saws, the sawdust, the builder stopping to explain his craft to anyone who paused. By the time Santa arrived he had a home, and the log builder had a month of the best advertising he'd ever had.

And I needed someone to run it. I went to Habitat for Humanity — a cause near to me, because I had known what it was to not quite have enough to secure a place to live. Their build schedule was full that year, but they had a new program for teenagers. So we built a program together: each evening, teenagers and their parents would manage the Santa set, and the money from the photographs came in as donations.

Then the cold handed us a gift disguised as a disaster. It was so cold that first year that the instant photographs wouldn't develop.

At first it looked like a failure. Instead it became another touchpoint. Parents had their picture taken, got a receipt, and had to come back the next day to the photographer's little studio to collect it — where he had one more chance to serve them, and they had one more reason to return to Birkdale Village.

That first year, the proceeds built three homes. All I really paid for were the candy canes.

I have never forgotten what that taught me: when you start looking for the connections, one touchpoint leads to another. The Santa program was never only about Santa. It tied Birkdale Village to Habitat, Habitat to teenagers and their parents, the teenagers to a cause worth their Friday nights, the photographer to new customers, and the families to another reason to come back. The place was quietly growing relationships around itself.

That partnership paid me back once more, in a way I could never have planned. The teenagers who filled the village on weekend evenings had become their own kind of story. After I left Eastland, that mall had grown rough — there had been a near-riot, it had made the news — and now the media were hunting a follow-up on "teen trouble at shopping centers," and they'd heard Birkdale Village drew a crowd. I was already worried, though not about a riot. I was worried about eleven-year-olds dropped at the curb for a movie, behind a pretty streetscape that hid four large parking garages. I had three daughters. I knew exactly what I was looking at. So I had handled it my way — two women walking the neighborhoods, leaving a heartfelt letter at hundreds of doors, reminding parents that this was a shopping center on a major corridor, not a small town, and asking them to be our partners in keeping their children safe. When the cameras finally arrived on a Friday night in December, I had a story to tell: the neighborhood letters, the added security, and teenagers who weren't loitering but running the Santa line for Habitat. The cameras turned toward those kids taking donations under the tree, and the trouble story became a story about belonging. I had built that partnership to staff a Santa house. It ended up protecting the whole place.

The Fourth of July began, as many things did, with Peter. He wanted a Nantucket-style water fight on Main Street. I had no idea how to *make* a water fight, and around here the Fourth belonged to the lake. So I called Nantucket. After a few transfers and a few chuckles, a man explained that they simply rolled out two fire trucks, unwrapped the hoses, and turned on the water. Everyone got soaked. That was the whole point.

I had two towns with fire departments. What I didn't have was a reason to get soaked — you need more than water; you need a contest. The night before, I stood in a Home Depot working out how much wire it would take to string a ball between two fire trucks so the spray would drive it back and forth. I bought a big gold trophy. Game on. The first bike parade was just me holding a banner in front of a few dozen kids on bikes and babies in strollers, on a street with no spectators because everyone was *in* the parade. It grew into thousands and it still runs today.

But I had, as I said, spent the budget on a Christmas tree. Everything else I had to fund some other way — and entertainment was central, because I needed people to know that on a Friday or Saturday evening there would always be something to do at Birkdale Village. The restaurants and shops were counting on it. Keeping them alive was my job.

So I went back to the question I have used my whole career to solve a marketing problem without money: *Who else needs to talk to the same people I am talking to?*

A new hospital nearby had our exact problem — people still drove into Charlotte to hospitals they trusted, and Lake Norman Regional was a stranger to them. So every weekend night, the banner over our stage announced that evening's performance sponsored by the hospital; its logo rode our printed program and our website. The hospital wasn't buying an ad. It was becoming part of the experience — becoming, week by week, a little more familiar, a little more trusted. There were two sidewalk spaces flanking the fountain; I gave one to a car dealer and one to a boat dealer, and told them to imagine what it was worth to be seen by a parent who was relaxed, watching a child play in the water, with music in the air. Those partnerships brought in real money, and taught me a lesson I have never let go of: marketing rarely needs more budget. It needs you to look differently at the people who could benefit from what you are already building.

And then there was the night that told me what I had actually made.

I had worked late. It was nearly ten, bitterly cold, and the village had emptied — the snow-watchers gone home, Santa's line closed an hour before. But the tree lights were still glistening and the music was still playing, and as I left my office I glanced toward the arbors and saw two figures. A couple, alone, holding each other and dancing slowly beneath the tree, under the stars, in the falling snow. No crowd. No Santa. No line. No one there to see them but me — and I don't think they knew even that.

I stopped. And somehow the emptiness made it more beautiful, not less.

I had spent years thinking about how to get people to come, what to show them, what to say, how to give them a reason to return. I had never once considered that the truest measure of all of it might be something that would never appear on any report I would ever hand anyone. That it might simply be the moment when two people felt so at home in a place that they stayed after everyone else had gone, and danced.

I had built the tree, and the lights, and the snow, and the music. I had created the conditions. But I had not created that moment. They had.

That is the closest thing I have to proof of everything this book argues.

I started with a thirty-thousand-dollar Christmas tree and two snow machines. What I was actually trying to build was harder to put on a budget — a reason to care, a place people could carry in their lives, a small town's worth of belonging assembled out of snow machines and gold trophies and log cabins and one impossibly good boss who let me try. And what I learned at Birkdale Village, and have spent the rest of my life turning over, is this: belonging can be built on purpose. You can lay every stone of the experience with intention — the snow, the music, the light. But the meaning is not yours to lay. You build the conditions. The people build the moment. And if you have built well enough, they will make something inside your careful little town that you could never have designed — and would not trade for anything on the ledger.

PART THREE

# What We Keep

CHAPTER 12

# What We Keep

*We do not hold onto moments because they were perfect.*
*We hold onto them because they meant something.*

I AM, BY LONG PRACTICE, THE QUEEN OF GARAGE SALES. Move fifty times in a life and you learn an art most people never have reason to acquire: the art of traveling light. You learn not to grow too attached to things, because things have to be packed, and carried, and found a place in the next house, and the next, and the one after that. Somewhere along the way I stopped experiencing this as loss and started experiencing it as a strange kind of freedom — one I have come to love. If someone admires something of mine, I am happy to give it to them. If an object brought me joy, let it go bring that same joy to someone else. I have held thousands of small farewells over folding tables in a thousand driveways, and I have rarely missed what I let go.

So I am, on its face, the last person who should write a chapter about the things we keep. I keep almost nothing. I have made an entire life out of releasing what other people cling to.

And yet.

There are a few things I have never been able to put on the table. Not many — they would fit in a single small box. A handful of photographs. A few pieces of artwork my children made me when their hands were still small. A strand of pearls from Donna. A tarnished silver bowl with my mother's initials on it. And a set of sleigh bells. To anyone wandering through one of my yard sales, it would look like the leftover hodgepodge at the end of the day — the odds and ends nobody bought, the things you finally box up and donate just so you don't have to decide where they go next. But these are the opposite of leftovers. In a life built on letting go, these are the few things that letting go could never reach.

I have learned something from that small box that I could not have learned any other way. When you can release almost everything and lose nothing essential, the few things you cannot release tell you, with absolute clarity, what was essential all along.

The photographs and the children's artwork are easy to explain and impossible to part with — they are my daughters, the people every hard mile of my life was ever for. The pieces are objectively nothing: paper, clay, color. They are also the most valuable things I own, because they are proof of who we were to each other before any of us knew how the story would turn out.

The strand of pearls came from Donna — the mentor who found me as she was on her knees planting silk flowers and saw something I could not yet see in myself. I keep them not because pearls are precious but because hers were given by a person who believed in me before I had earned it. To hold them is to be seen again, the way she saw me. Some objects are just being-seen made into something you can carry.

The silver bowl is my mother's, her initials worn into it, the surface gone soft and dark with tarnish I have never polished away and never will. She is the one who taught me to travel light in the first place — who framed every uprooting as an adventure, who could make leaving feel like possibility. There is a quiet irony I have made my peace with: of all she gave me, the lesson to hold things loosely was the greatest, and the one object I could not hold loosely is hers. The tarnish is not neglect. It is the years themselves, and I would not trade them for shine.

And then there are the bells.

A set of sleigh bells, kept and carried through every move, every house, every driveway sale where everything else found a new home. I have given away furniture, dishes, books, art, clothing, the accumulated weight of decades — and somehow the bells always went back in the box that came with me. For a long time I did not ask myself why. They are not valuable. They are not even particularly beautiful. But I think, now, that my hands knew something before I did.

When I was a child at Chautauqua, I never needed a watch, because the bells in the tower rang the time across the rooftops and out over the lake, and the hour simply came to me wherever I was. Time, in that place, was not something I monitored. It was something I received — musical, unhurried, a part of the very air. I spent most of the years that followed forgetting that lesson, living out ahead of every moment, checking the time instead of inhabiting it, until a night in the desert stopped me long enough to remember what it felt like when time deepened and held instead of racing past.

I think I kept the bells because they are that memory made into something I can hold. When I pick them up, they still ring — the same clear, insistent sound, the sound of winter and arrival and Christmas mornings I fought to make magic for my girls.

And in that sound the whole of my life gathers for a moment: the tower across the water, the child who did not yet need a watch, the woman who learned the hard way that a life can either slip past unnoticed or be marked, heard, felt. The bells are my refusal to let it slip past. Of everything I own, they are the one object that insists time does not have to be lost. It can be rung.

This is what the things we keep are really for. Not the objects themselves — I, of all people, know how little the objects matter, how freely almost all of them can go.

What we keep is never really the thing. It is the person inside the thing, the season inside it, the time we refused to let pass unmarked. A keepsake is a marker of belonging. A ritual is a way of returning to ourselves. They are technologies of memory — the way human beings have always slowed time down enough to feel it, and held onto what would otherwise blur together. And the deepest lesson of my small box is this: I could let go of nearly everything and lose nothing essential, because the few things worth keeping were never things at all. They were people. They were time made audible.

I have moved more than fifty times, and I have told you that I keep almost nothing. But there is one thing I have carried through every single move — the one thing I never once had to pack. It took me most of my life to recognize it, and longer still to say out loud.

For decades I searched for home the way you search for a set of lost keys, certain it was a particular place waiting out there somewhere, if only I could find the right town, the right house, the right life. I chased the feeling Chautauqua had given me. I looked for it in every place I landed, and again and again I concluded that I had not yet arrived.

What I could not see until I sat down to write this book is that home was never a place at all. It is a space within me. It is the quiet room I step into on the rare occasions I stop racing ahead of my own life and actually arrive in the moment I am standing in. A night in the desert would eventually show me the door to that room — I will come to it — but the room itself had been there all along, furnished and waiting, in every house and every driveway sale and every unfamiliar town. I had simply been looking out, when the whole time I should have been looking in.

That may be the deepest thing I have ever kept. Not the bells, not the pearls, not the tarnished bowl, precious as they are — but the slow-dawning knowledge that I could be at home anywhere, because home was something I carried inside me. And if you have spent your life, as I did, hunting for the right address, let me tell you gently what took me the better part of a lifetime to learn: you can stop searching out there. The place you are looking for is a space within you, and there are ways — small, repeatable, and entirely within your reach — to find your way back into it. This whole book is, in a sense, a map to that room.

I spent a career, it turns out, helping companies and places do on purpose what those five objects do for me — hold meaning in place so it does not slip away. A brand at its best is a kind of keepsake: a thing people carry because of what it holds. The packaging that feels like a gift, the ritual that becomes a tradition, the small, repeated gesture that says you belong here — these are touchpoints, and they are how an experience becomes something people refuse to put on the table and sell. But that is the work of the chapters ahead. For now, this is just a tarnished bowl, a strand of pearls, a child's craft, and a set of bells that still ring — and everything they will not let me forget.

*We keep objects and rituals because they hold our memories in place. But memory does not always honor the things we expect it to. Sometimes the hardest season of a life is remembered, by the people who lived through it beside us, as something else entirely — as an adventure, even as joy. I learned this from my own daughters, years after a journey I had spent terrified that I was failing them. The next chapter is the trip the way they remember it.*

CHAPTER 13

# The Courage to Have No Plan

*Sometimes we are most alive in the very moment we feared.*

We return not only to places, but to feelings we hope to find again. And the feelings we return to are rarely the polished ones. Some of the most memorable experiences of my life were not polished at all — they unfolded in the middle of uncertainty and exhaustion and complete unpredictability, and were remembered, afterward, as the best of times. I have come to believe this is one of the most important and least understood truths about being human, and I learned it, like most things, the hard way.

Years ago, during one of the most difficult seasons of my life, I packed up my daughters and drove from one side of the country to the other. We had spent a stretch living with family through an enormous transition; we had sold nearly everything we owned; and our future felt like a question none of us could answer. I had very little money, no cell phones, and no carefully structured plan

— beyond one fierce determination: somehow this had to feel like an adventure for my girls, and not like a collapse.

At the time, I genuinely believed I was creating the ultimate road trip. Looking back, I can see how wildly questionable much of my planning actually was. My oldest had just graduated from high school and was driving her own small, ancient two-seater behind me — through mountains, deserts, tiny towns, and long empty stretches where nothing existed but heat and distance. I had maps, camping gear, a bit of cash from selling our belongings, and a level of optimism that probably should have worried someone. Still, we went.

We drove through the redwoods, those impossible cathedrals of trees. We followed the California coast. We saw Hollywood, pointed the cars toward the border, and then turned back east through Arizona. It was beautiful and strange and entirely ours. The girls picked the music. We ate badly and laughed often. We were, for those weeks, a small self-contained world moving across an enormous country, and whatever waited for us at the other end, we were in it together.

Somewhere near Phoenix, one of the cars began to overheat, the way old cars do in desert heat — running fine until it didn't, then refusing to start again until it had cooled. My solution, naturally, was optimism mixed with improvisation. We stopped for a day at a swimming hole near Flagstaff, and I quietly calculated that when the car inevitably failed afterward, I would call for a tow, get us the twenty miles into Flagstaff, find somewhere to spend the night, and figure out the rest from there.

That was the plan. Life had another one.

The tow truck arrived, and the driver informed me he would not be taking us to Flagstaff after all. He needed to head back toward Phoenix — more than two hours the wrong way — because he had somewhere important to be that evening. I did not exactly have better options, so I agreed, and we loaded up and set off into the dark in the wrong direction. At some point he mentioned he needed to stop by his house to change clothes before he continued on. He disappeared inside for a while. And he came back out dressed as a clown.

Not metaphorically. Literally. He was a rodeo clown, due at a rodeo that night, and the costume was simply his work clothes. Even then, somehow, this did not strike me as nearly as strange as it should have. By that point in the trip, a clown towing my dead car through the Arizona night seemed about right.

By the time we reached the rest stop above Phoenix, night had fallen and the heat had finally broken. The clown lowered the car off his lift and told me it would run fine now — the engine only stalled in the heat, he said, and the heat was gone. He climbed back into his truck and drove off to wherever clowns go at night, and the sound of him faded, and then there was nothing.

I stood in the dark in the middle of the Arizona desert. My three daughters were in the cars, tired, trusting, certain their mother had a plan. There were signs posted at the edge of the lot warning travelers to watch for rattlesnakes. I had no cell phone. I had no credit cards. I had a little cash from selling almost everything we owned, two cars held together by optimism, and several hundred miles of darkness in every direction.

And I cried. Just a couple of tears — I could not afford more than that, and the girls could not see them. Because standing there, I knew something I had been outrunning since California. I had done this. No one had done it to me. I had put my children at risk, alone in the dark with the snakes and the distance, and I had done it chasing something good. I had wanted to give them an adventure instead of a collapse. I had wanted them to remember the redwoods, the ocean, the glamour of Hollywood — instead of everything we had lost. I had been so busy building the memory that I had driven us straight into the danger.

That is the version I never told them. For years the story was just the clown — the absurd, wonderful rodeo clown — and the way we laughed about him afterward. That was the memory I built, and I built it well; one of my daughters told me, decades later, that she never knew how serious our situation had been. That was my job: to protect them, to make sure the danger stayed mine and the adventure stayed theirs.

I think a great many of us are doing some version of this — holding a fear we will not set down in front of the people we are protecting. There is courage in it. There is also a particular loneliness, the loneliness of being the only one who knows how close the edge really was. If you have ever been the one keeping the fear off everyone else's face, you know the exact weight I am describing. It does not photograph. It does not make it into the story anyone else gets to keep. But it is real, and it is heavy, and it is carried mostly in the dark.

But there was another thing I learned that night, standing between the dead car and the rattlesnake signs, and it took me most of my life to understand it. I had spent every mile of that trip — every mile of those years — three steps ahead, planning the next move, scanning the road for what could go wrong, because that is what you learn to do when you have lived in survival mode. You leave the moment, because the moment is where the fear lives.

And yet that was the most alive I had ever been.

No plan. No map out. No one coming. Just me and my girls and the enormous desert sky, and the strange, terrible, exhilarating freedom of having no idea what came next. It takes a particular kind of courage to be in the moment — I know, because it is the one form of courage I have always lacked.

Back in Maine, years before, I had learned to live ahead of my own life; I had bought three pairs of skates for a dream that was mine and missed the guitar my daughter was quietly asking for, and I had decided the lesson was that I should have planned better. That night in the desert taught me the opposite. The planning had finally run out. There was nothing ahead to scan for that I could do anything about. There was only the dark, and my children, and the next hour. I had to simply be there.

And being there, with everything stripped away and nothing certain, I understood for the first time that I was not getting *through* the greatest adventure of my life. I was finally *inside* it.

The time was not racing past me, the way it always had while I lived out ahead of it. For one night it stopped, and deepened, and held — the way the bells used to make time hold when I was a child who did not yet need a watch, because the hour simply came to me across the water. I had spent decades checking the time. That night, for once, I lived inside it.

I would love to tell you the lesson took, all at once — that I drove out of that desert cured of living in the future. It did not. I am still, to this day, a planner who has to be reminded, sometimes daily, to come back into the room I am actually standing in. But I have carried that night with me ever since, the way we carry the things that cost us something. It is why I notice, now, when I have left my own life without moving from my chair. It is the truest thing I know, and the hardest to practice: that being ready is not nearly as important as being real, and that the moment we are so frightened of — the one with no plan, no map, no guarantee — may be the only place we are ever fully alive.

That is what I kept from the desert.

Here is what my daughters kept.

Years later, I was driving alone in Charlotte, listening to a radio program where listeners call in to tell their stories. And my daughter's voice came over the air — a recording made some time before — telling the story of the rodeo clown. She was laughing as she described our bizarre cross-country adventure and the tow-truck driver who turned out to be on his way to a rodeo. I pulled the story toward me like something warm. And in that moment I understood something I have never forgotten.

What my daughters remembered from that season was not the fear. Not the uncertainty. Not the money we did not have, or the life we had sold off, or the loss of everything familiar. What they remembered was the adventure. Being together. Figuring it out as we went. The laughing. The absurd and colorful rodeo clown. Their touchpoints were not the hardship. Their touchpoints were the feeling that surrounded it.

Here is the truth I most want you to carry out of this chapter, because it took me a lifetime to see it and it changes everything about the work I would spend my life doing. We do not remember our lives according to the scale of an event. We remember according to emotional texture — according to connection, according to whether we were fully alive inside the experience. The polished moments are not always the ones that stay. Sometimes what stays is the breakdown, the detour, the impossible night that turned, against all odds, into a story we tell for the rest of our lives.

I had built that season for them on purpose — the laughter, the adventure, the same way I once bought three pairs of skates and called hardship by another name. That was the work of a mother, and I would do all of it again. But here is what the desert finally taught me, and what the radio taught me years after: I could author the conditions of that trip. I could not author what my children would make of it. I gave them a broken-down car and a rattlesnake-warning sign and a mother frightened half to death — and out of it they built an adventure they would carry, laughing, into their own adulthoods.

You can lay every stone of an experience with intention. What the other person keeps is finally theirs to build. And if you have loved them well enough in the building, they may make something out of your hardest night that outlasts the hardness itself.

CHAPTER 14

# The True Rattlesnake

*There comes a time — for a business, a brand, or a person — when you cannot make enough magic to hide a foundation that will not hold.*

One of my daughters laughed when she read the chapter about the rodeo clown and the desert. She loved it — the tow truck, the man dressed for a rodeo, the whole improbable adventure. But she said I had left something out. "Mom," she said, "you never mentioned the real rattlesnake."

She was right. There was one on that trip — an actual rattlesnake in the dark, one more danger in a night already full of them. But she meant more than that, and we both knew it. The fear of that night was real, and so, many times over the years, was the flight. But there was a longer danger underneath the desert one, and I have not told you about it yet, because I did not know how to tell it without letting it swallow the rest.

I was married for twenty-five years. I was a "for better or for worse" kind of woman, and I held that promise close, long past the point when most people would have set it down. I have always said I take responsibility for nearly everything that has happened in my life — and I mean it, because the same grace lets me take responsibility for all the good, too. But the truth I have been circling this whole book is this: the "worse" was bad almost from the beginning. Bad enough that, for a very long time, I was the only person left who still believed I could make it work.

That is what all the magic was for. I told you early on that I spent a career building belonging for strangers because I could not find it for myself. This is the part I left out: I was building it at home, too, day after day, trying to make a place feel like something it was not. I was very good at it. But we all reach a moment — in a business, a friendship, a family, a marriage — when we cannot make enough magic to hide a foundation that will not carry the weight of what we owe the people we love, or ourselves.

In business, this has a name and a remedy. When a brand is tarnished, you rebrand. You drop the product, you change the personnel, you issue a statement and move on. A life is not so clean. There is no press release — only the slow, private accounting of what is true.

I will not name names, because this is not a chapter about blame. I chose my own rattlesnake, and I stayed a long time; those were my choices, and I own them. This is a chapter about the moment the accounting finally came due — and what I did with it, which is the only part that might be of use to you.

Here is what I did not do: I did not make a plan. That may sound strange coming from me — the woman who lived in perpetual planning mode, who could engineer a spectacle for ten thousand people. But there was no strategy, no five-year plan, no spreadsheet for this. One ordinary morning I simply woke up and thought, with a clarity I had never let myself feel: I need to make a change. That was all. And it is the most honest thing I can tell you, because it is the whole argument of this book arriving in the one place I least expected it. The change did not come from being ready. It came from being present — from standing, at last, inside my own life instead of out ahead of it, and seeing it plainly.

The path after that was mine to walk, and mine alone. It asked more of me than I believed I had. It meant finding a new identity while carrying my whole history with me, unerased.

So let me say to you what I most needed to hear then — whether you are a person, a place, or a brand standing at the edge of a foundation that will not hold. You may not believe you have the capacity to change course. You do. You don't have to have a plan to begin; you need only to be present enough to tell yourself the truth, and brave enough to take the first step before you can see the second. Sit in the moment. Ask for the wisdom of the people who love you. The way has a way of appearing.

I have never regretted it. Not once.

*The desert taught me that presence could be survived. This taught me it could save me — and that the same courage it takes to change a life is the courage it takes to build anything real. What breaks can also be rebuilt, if we are willing to stand present in the ruins long enough to see the way.*

PART FOUR

# Designing Experience

CHAPTER 15

# The Details That Speak

*Every detail is speaking. The only question is whether anyone decided what it should say.*

UNTIL NOW, I HAVE TOLD YOU ABOUT THE TOUCHPOINTS life handed me. The skates and the guitar. The bells across the water. A mentor on her knees in a mall, planting silk flowers into a life she could see before I could. A dead car in the desert and the one night I finally stopped running in fast-forward. Those were the moments that made me — the ones I did not design, and mostly did not choose.

Now I want to turn the book around.

Because there is another half to this story, and it is the half I actually got paid for. While I was struggling, for most of my life, to find presence for myself, I was building it, on purpose, for other people — for strangers who would never know my name, in places that began as nothing but bare ground. I spent a career engineering the exact thing I could not hold onto. The wound and the work were always the same subject. And having learned, in the hard years, what presence costs, I want to show you how I learned to build it.

So let me finally say what I am, by trade. What I did, for twenty-five years, is something I'd call a storyteller of place — someone who takes empty land, or a mall, or a new neighborhood, or a resort, or a shopping center that could have been anywhere, and turns it into *somewhere.* Somewhere people fold into their lives. Somewhere they return to without quite knowing why.

When my career began to truly blossom, it opened doors I had never imagined. As Director of Marketing at Eastland — and later at Birkdale Village, The Wharf, and in my work with Christie's International — I found myself in rooms I never dreamed I'd enter: flying on private jets, walking the red carpet at Mann's Chinese Theater for the premiere of the Spice Girls movie, hosting Beyoncé back when she was still with Destiny's Child, trading a few words with Elizabeth Taylor, being photographed with Joan Jett. There were others, too — Ted Danson and Mary Steenburgen, Tiffani Thiessen, Hank Williams Jr. — and every time, without exception, I was thrilled to be included and quietly amazed at how gracious they all were. It was such a juxtaposition with my early adult life that I sometimes had to pinch myself to believe how far I had come. But what struck me most, every time, was how deliberately their teams made every touchpoint count — how nothing was left to chance. That became one of the great influences on me: it sharpened how I thought about presentation, and deepened the passion at the heart of this book — for how we present ourselves, a place, a business, or a brand.

And the whole craft, it turns out, comes down to one deceptively small idea. Every detail is speaking. The only question is whether anyone decided what it should say.

Let me give you the word plainly, because I have been leaning on it and I owe you a definition. A *touchpoint* is not a transaction, and it is not a feature on a list. It is any moment through which a person receives a message about whether they matter. A greeting. A wrapped box. A returned call. A clean restroom. The pause before a welcome. That's the key — and it is very nearly everything.

Here is why a thing that small is worth a book. Touchpoints do not stay small. This is the chain that runs underneath everything I am going to tell you: a touchpoint creates an experience; experiences, felt and repeated, settle into memory; memory, over time, hardens into meaning; and meaning, in the end, becomes part of who we are. A wrapped box, a clean restroom, a returned call — none of them is small at all. They are the raw material of memory, and memory is the raw material of a self. That is why this book is called what it is. Every touchpoint is a small deposit into the person you are becoming, and into the person you are helping someone else become.

So I want to be precise, because this is the foundation everything else is built on. Experience is never neutral. It is never simply happening. It is always being authored — by intention or by neglect, but authored either way. The way a place greets you or fails to. The way a message lands or grates. The way a room is lit, a question is answered, a line is managed, a gift is wrapped. None of it is background. All of it is telling the person on the receiving end something true about whether they matter here.

We tend to think the message a business sends lives in its words — its slogans, its ads, the things it says about itself. It doesn't. The message lives in the touchpoints. It lives in the accumulated evidence of a thousand small moments, and people read that evidence fluently, instantly, mostly without noticing they are doing it. You have done it. You have walked into a place and known, in the first ten seconds, whether you were welcome — before a single word was spoken. That knowing did not come from nowhere. It came from details, and the details were speaking.

Think about the Apple box — I described it early in this book, the strange care poured into a container most people tear open and throw away. Look at it now with a builder's eye. None of that effort makes the phone work better. All of it is a sentence, delivered before you have read a word: *we thought about how this would feel for you.* That sentence is the first thing the company says, and it says it without speaking, and you believe it — because a detail cannot lie the way a slogan can. Anyone can claim they care. The box proves it.

Places speak in the same grammar, in a language most people never notice. Think about the last time you walked the Main Street of a well-made themed environment — a Disney park is the cleanest example. You felt pulled gently forward, toward the castle at the far end, and you probably never asked why. The why is that nearly everything you were looking at had been engineered to produce exactly that pull. The street narrows almost imperceptibly as it rises. The upper floors are built smaller than the lower ones, a forced perspective that makes the castle feel taller and farther and more storybook than it really is. A trash can is never more than a few steps away, so you never hold a candy wrapper long enough to feel irritated. The sightline is protected, so the thing you are meant to feel wonder toward is always framed and never cluttered. You experienced none of this as design. You experienced it as magic. That gap — between what you felt and what was done to make you feel it — is the entire discipline I gave my life to.

Even without a castle, or a Disney budget, you can speak this fluently. Sometimes the most eloquent touchpoint in the world is a bathroom.

Let me tell you about the unlikeliest cathedral of touchpoints I know. It is a gas station. If you've driven through Texas — or, increasingly, a dozen other states — you've seen them, or seen the billboards counting down the miles: Buc-ee's. A travel center the size of a small airport, a hundred-some fuel pumps, a beaver for a mascot. People don't stop at Buc-ee's because they need gas. They plan road trips around it. They buy the t-shirt. They will detour an hour to stand in a gas station.

How does that happen? The way everything in this book happens: one touchpoint at a time, starting with the humblest one imaginable. Because Buc-ee's was built on a bathroom. When Arch "Beaver" Aplin opened his first store in 1982, he made a decision that sounds absurd until you feel it yourself: the bathrooms would be immaculate. Not tolerable. Immaculate. He understood what his competitors never bothered to learn — that for a traveler, exhausted and far from home, a clean restroom isn't a convenience. It's a kindness. It is the difference between feeling like a nuisance and feeling like a guest. The company later won an award for the cleanest restrooms in America and has guarded the title ever since, employing people whose whole job is to keep them spotless, and paying them well to do it.

Start there, with the touchpoint everyone else treats as an afterthought, and watch what it makes possible. You cannot sell someone a beaver-shaped keychain in a filthy gas station. But in a spotless one, where you already feel cared for, you'll wander — and so they gave you somewhere to wander: brisket carved to order, homemade fudge, the caramel-corn confection people haul out by the bagful, souvenirs customers wear like a badge. A gas station became a *place* because someone decided every point of contact, down to the toilet, would say the same thing: *you matter here.* Buc-ee's didn't out-market anyone. It out-*cared* them, relentlessly, in a thousand small decisions, until caring became the brand. Nobody drives an hour out of their way for cheap gas. They drive it for the feeling — and the feeling was built, deliberately, beginning in the last place you'd think to look for magic.

I call this learning to see experience as architecture. Most people move through the built world receiving its signals without ever noticing the signals exist — the way you can love a song for years before you can name a single chord. The storyteller of place learns to hear the chords. To walk into a room and see not the furniture but the decisions. To recognize that the pause before a welcome, the tone of a subject line, the sightline from the front door, the way the light falls at four in the afternoon — all of it is language, and all of it is either saying something on purpose or saying something by accident.

The practice of it, early in my career, was almost embarrassingly simple, and I still recommend it to anyone building anything. I would walk into a place and try, as hard as I could, to arrive as a stranger — to forget everything I knew and simply feel what a first-time visitor would feel, step by step, from the parking lot in. Where did my eye go first, and was that where they wanted it to go? What did the entrance ask of me — did it welcome me, or make me work to find the door? What did the place smell like in the first breath? Where did I feel lost, or rushed, or unsure whether I belonged? Every one of those reactions was a touchpoint reporting for duty, and most of them, in most places, had never been decided by anyone. My whole job was to walk in, hear what the place was accidentally saying, and then, one detail at a time, change the sentence — move the entrance, warm the light, open the sightline, slow the welcome — until the accident became an intention.

And here is the truth whether or not you ever build anything larger than a dinner table. You are already fluent in this language. You have been reading touchpoints your whole life — that is what being human in a designed world trains you to do. What you may not have realized is that you are also *writing* them. Constantly. In your home, your work, your gatherings, the small daily exchanges that make up most of a life, you are authoring an experience for the people around you, and every detail of it is speaking on your behalf. The email you send at eleven at night. The way your entryway greets a guest, or doesn't. Whether you look up when someone walks into the room.

Try, once, the stranger's walk on your own front door. Come up the path as a guest would. What does the entrance say — is it warm, or is it a pile of shoes and a burned-out bulb and a note of apology waiting to be spoken? I am not scolding; my own doorways have said unlovely things for years at a stretch. I am pointing at the fact that they were saying something, and that I had not decided what. A home is a set of touchpoints exactly the way a hotel is. So is a marriage. So is the way you answer the phone when your mother calls.

Because we are always deciding, even when we think we are not. What I want you to carry out of this chapter is this truth, There is no neutral. A detail left to chance is not silent; it simply says something you did not choose. The unanswered message says *you did not matter enough to answer.* The cold room says *come in, but not too far.* The rushed goodbye says *I have already moved on to whatever is next.* None of those was anyone's intention. That is exactly the point. Neglect has a voice too, and it is often the loudest one in the room.

I learned to build by learning to listen for that voice — to walk a property and hear everything it was accidentally saying, and then, one detail at a time, to change the sentence. That is the work. Not decoration. Not marketing. The slow, deliberate authorship of what a place, or a person, or a business tells people about whether they belong.

It begins with a single act, and it is the same act whether you are building a town square or setting a table for people you love: you decide the details will speak on purpose. And you remember, as you decide, what the small thing in your hands actually is — a moment through which someone will learn whether they matter, and keep, long after they've forgotten the day, the feeling it left behind.

CHAPTER 16

# What Attention Says

*Trust is built slowly and broken quickly. In the end, attention is the whole currency.*

TRUST IS BUILT SLOWLY AND BROKEN QUICKLY. ANYONE who has ever loved another person knows this in their body. You can spend years being reliable, and lose a great deal of it in an afternoon. The same law governs places and brands and businesses, and it is the first hard thing anyone who wants to build experience has to make peace with: the touchpoints are not weighted evenly. A hundred small kindnesses can be quietly erased by one careless moment, because the careless moment is the one people remember.

I want to sit with that asymmetry, because it is easy to hear it as unfair and miss what it is actually telling us.

We are wired to notice the break in the pattern. It is old wiring, older than commerce, from a time when the anomaly in the landscape was the thing that could kill you. A thousand safe walks through the same field teach you nothing you can feel; the one time something moves in the grass, you remember forever. So it is with experience. When a place has told you, through a hundred small signals, that you matter here, you stop consciously noticing the signals — they become the air. And then one touchpoint contradicts all the others, and suddenly you notice everything. The spell does not fade. It snaps. And once it snaps, you begin, unfairly but unavoidably, to reread all the earlier signals in its light, wondering whether the care you felt was ever real.

This is why neglect is so much more dangerous than most people building anything ever realize. They imagine the threat is a dramatic failure — the ruined meal, the lost order, the public mistake. Sometimes it is. But far more often, trust dies from something much smaller and much quieter: the touchpoint nobody decided about. The follow-up that never came. The sign left up three weeks after the sale ended. The voice that went flat on the fiftieth call of the day. The email answered in a tone that was technically polite and entirely absent. None of these are disasters. Nobody meant anything by them. That is precisely why they are so corrosive — because they tell the truth. A place that stops attending to its details is telling you, accurately, that its attention has moved on. And people can feel the difference between being cared for and being processed as surely as they can feel a change in the weather.

Let me give you the clearest test I know, because it is one you can run yourself the next time you are deciding whether a place's care is real or merely performed. Go find the restroom. A business will lavish attention on the touchpoints it knows you are judging — the lobby, the storefront, the menu, the greeting. Those are the parts it expects to be graded on, so those are the parts that get the effort. The restroom is different. The restroom is the touchpoint a business assumes you will not use to form an opinion, which is exactly why it tells the truth.

A restaurant with a warm host and a beautiful dining room and a neglected, tired, faintly unpleasant restroom is telling you something it does not know it is telling you: that its care extends precisely as far as the places it is being watched, and no farther. The care is a performance, not a character. And some part of you registers that, and quietly recalibrates how much to trust everything else. The details a business thinks don't count are the ones that reveal what it actually is.

I have been on the receiving end of neglect, the way we all have, and it lands harder than the size of the thing would ever justify. You walk into somewhere you once loved, and something has gone slack. You cannot always name it — the same logo, the same layout, the same product — but the feeling that used to be there has thinned. What changed was never announced. What changed was that someone stopped deciding, and the details began saying what neglected details always say: this used to matter to us more than it does now. It is one of the loneliest messages a place can send, and it is sent, every day, entirely by accident. It is the commercial cousin of a feeling I have known in my own life — the particular ache of standing in front of someone and being processed rather than seen, handled efficiently and not quite met. We are hungry to be seen, all of us, and neglect is the precise opposite of being seen. It is being confirmed as not worth the attention.

Here is the part that should give us hope, though, because it is genuinely hopeful once you see it. If neglect is a failure of attention, then repair is almost never a matter of money. It is a matter of noticing, and of giving people permission to act on what they notice.

Consider how the Ritz-Carlton handles the moment when something goes wrong. Famously, the company authorizes every employee — the housekeeper, the front-desk clerk, the person clearing plates — to spend up to two thousand dollars, per guest, to resolve a problem, without asking a manager. Most people hear that number and fixate on it, as though the point were the money. The point is not the money. Almost no one ever spends anywhere near it. The point is what the policy says to the employee, and through the employee to the guest. It says: you are trusted to see the person in front of you, and you are trusted to act. It converts a rule-following clerk into an attentive human being with the authority to matter. The guest who has a small disaster smoothed away by someone who simply decided to fix it does not remember a transaction. They remember being caught when they fell. And they tell that story for years.

Notice what actually happened there. The failure — the broken touchpoint — did not destroy the trust. Handled well, it deepened it. This is so reliable that it has a name; people who study service call it the recovery paradox — the counterintuitive fact that a customer whose problem is resolved beautifully often ends up more loyal than one who never had a problem at all. It sounds impossible until you remember how trust actually works in human beings. We do not fully trust the person, or the place, that has never been tested. Anyone can be gracious when everything is going right.

What we are really looking for, underneath, is evidence of how we will be treated when things go wrong — because that is when it counts, and that is when most people and most places look away.

The one that does not look away, that reaches out a hand exactly when a lesser one would have retreated behind a policy, earns something a flawless record never could. It earns the trust that only survives contact with trouble.

So neglect and recovery are two sides of the same coin, and the coin is attention. The neglected touchpoint says our care has moved on. The recovered one says we are still here, and we saw you, and we will not let this be the thing you remember. The difference between them is not budget or brilliance or a bigger staff. The difference is whether the people closest to the moment are paying attention and have been trusted to act.

And this, too, scales all the way down. You don't have to have a two-thousand-dollar policy. You need the willingness to notice when a thing has gone slack in your own life — the friend you have stopped really answering, the ritual you have let thin, the room in your house that greets people with clutter and apology. Those are neglected touchpoints, and they are saying something you did not choose. Think of the friendship that did not end in any dramatic way but simply went quiet, one unreturned message at a time, until the silence became the relationship. No one decided that. That is exactly how it happens.

The repair is the same as the Ritz's, and just as cheap: notice, and act, before the person on the other end concludes that your attention has moved on. Most of the damage we do to the people and places we care about is not cruelty. It is inattention. And most of it is reversible, the moment we decide to look.

I spent years teaching crews and clients this one idea, usually against their instinct to chase the grand gesture. The grand gesture is not where trust lives. Trust lives in whether the small things are tended — reliably, repeatedly, especially on the ordinary days when no one is watching and it would be so easy to let one detail go. Because someone is always watching. Someone is always reading the evidence. And the evidence of neglect is the easiest sentence in the world to write by accident, and the hardest to take back.

*We have looked now at the shadow side — how easily a touchpoint breaks, and how the recovery can matter even more than the thing that broke. Underneath all of it sits a single principle: attention is the whole currency. So the real work is learning to spend it well, and the surprising truth is that the most powerful place to spend it comes before the experience even begins — before the guest arrives, before the customer walks in, before the first word is exchanged. The next chapter is about the touchpoint almost everyone forgets: the one that does its work in advance.*

CHAPTER 17

# Before You Arrive

*The experience begins before anyone arrives.*
*Anticipation is a touchpoint we either design or neglect.*

THE EXPERIENCE BEGINS BEFORE THE PERSON ARRIVES. This is the touchpoint almost everyone forgets, and it may be the most powerful one of all, because it does its work in advance — in the imagination, where there is no clutter yet, no line, no disappointment, only anticipation. By the time a guest walks through your door, the experience is not beginning. It is continuing. The only question is whether you authored the part that already happened, or left it to chance.

We underestimate anticipation because we cannot see it. But it is doing enormous work, and the work it does is not merely to make people excited. It is to set the lens. The mind does not wait for the body to arrive; it starts the experience the moment it hears about it, building a picture, forming an expectation, deciding in advance how to feel. And that expectation quietly bends everything that follows. The same glass of wine tastes better when we have been told it is expensive. The same meal satisfies more when we have looked forward to it for a week. The same room feels warmer when someone told us, before we came, that they could not wait to have us there. This is not gullibility. It is how perception works. We do not experience reality directly; we experience it through the frame we bring, and the frame is built in advance. Prime a person to expect warmth and they will find the warmth in an ambiguous moment. Prime them to expect indifference and they will find that instead. Whoever sets the frame has already half-written the experience.

Here's what it looks like when someone understands this, because I want you to feel the difference between information and invitation.

In the winter of 2020 my family planned a trip to a place called The Ranch at Rock Creek, a guest ranch tucked into the Montana wilderness near Philipsburg. We would not arrive for weeks. And yet the experience had already begun, because of the way the ranch chose to tell its story before we ever set foot on the property. Their website was not simply informational. It was immersive — beautifully produced films, rich photography, editorial storytelling that did not describe the ranch so much as let you stand inside it. And it did something clever and generous: it showed the place in every season, so that no matter when you were dreaming of going, the dream was already vivid. Winter offered sleigh rides and horseback trails through snow-hushed forests and the glow of lodge fireplaces. Summer promised fly fishing in cold clear streams and evenings under an enormous field of stars. Fall burned gold with aspens. Spring woke the whole landscape back up.

Here is what that website was actually doing. It was not selling accommodations. It was building anticipation — layering a feeling into us, weeks in advance, so that by the time we drove through the gate we were not strangers arriving at a hotel. We were people returning to a place we had already fallen a little in love with. The ranch had reached us before we arrived. That is the pre-arrival touchpoint at its finest: not a transaction confirmed, but a relationship begun in the imagination. And notice the discipline in it — the site could easily have been a list of amenities and a booking button. Instead, someone decided that the first job of the website was not to inform but to make us feel, and to make us feel it early.

Think about how differently that lands than the way most places treat the before. Most treat it as logistics — a confirmation number, a map, a list of policies. All necessary. All silent about how it will feel. And so the guest arrives cold, the experience beginning from zero at the door, every good thing about to happen having to overcome a blank or anxious frame instead of riding an eager one. The difference is not in the destination. It is in whether anyone bothered to author the anticipation.

You have felt the smallest, purest version of this in a darkened theater. In the last thirty seconds before a performance, the house lights dim — not all at once, but a slow fade, and a hush moves across the room without anyone requesting it. Nothing has happened yet. The curtain has not moved. But your body has been told, through a single wordless touchpoint, that something is about to matter, and you have made yourself ready to receive it. That dimming is pure pre-arrival design. It manufactures reverence in the seconds before the thing it is preparing you for. A great host does with a doorway what the theater does with its lights, and a great brand does it with everything a person encounters on the way in.

I did a version of this once, years before I had words for it, for a place that did not yet exist. I told you earlier how I stood in front of the developer at Birkdale — a Nantucket-style village that was still bare dirt and framing — and promised him I could make it snow on that main street. What I was really proposing, though I could not have said it this way then, was to build the anticipation before we built the place: to begin the experience in people's imaginations while the concrete was still wet, so that when they finally came, they were not evaluating a construction site. They were arriving somewhere they had already been promised, already pictured, already half-loved. Long before a single family moved in, the story had gone ahead of the place, and the place had only to keep the promise the story made.

The truth about the "before" is that it reaches past business into the shape of an ordinary life.

Anticipation is itself a form of presence. To look forward to something is to be reached by it early, to let it into your days before it technically arrives. We know this instinctively as children — the ache of the week before a birthday, the countdown to a trip, the way Christmas Eve can hold more magic than Christmas Day. Somewhere in adulthood we let that go. We stop building anticipation, for others and for ourselves, and in doing so we quietly flatten our own experience, letting things arrive unannounced and therefore only half-felt. A life with no anticipation in it is a life that is always starting from zero, never leaning forward into anything.

The people who make life feel rich have not forgotten this. They are the ones who set the table the night before, who send the note that says I cannot wait to see you, who describe the trip in loving detail long before the bags are packed. They understand, without needing the theory, that the invitation is part of the gift — that how you prepare someone to arrive is already shaping what they will feel once they do. You can do this for the people you love with almost no resources at all. A text that builds the moment instead of merely confirming it. A meal announced with a little ceremony. A welcome that begins before the doorbell. The dimming of your own house lights, in whatever form that takes, so the people walking in feel that something here is about to matter. Every one of those is a pre-arrival touchpoint, and every one of them tells the person the same true thing the best places tell their guests: I was thinking about this — about you — before you got here.

Because that, in the end, is what anticipation says. It says you were on my mind in your absence. And there are few messages a human being is hungrier to receive.

*The ranch reached us before we arrived, and it set a frame we would spend the whole trip living inside. But a frame is only a promise, and a promise made in advance is a debt that comes due at the door. The hardest, most revealing work of experience is not the anticipation. It is what happens when the guest finally arrives and the place has to make good — day after day, detail after detail, in a thousand small moments no website can stage. That is where I want to take you next, back to Montana, into the fully built world.*

CHAPTER 18

# The New Storytellers

*Storytelling no longer belongs to a specialized few. Every person and every place is telling one now.*

THERE WAS A TIME WHEN STORYTELLING BELONGED TO A small and specialized few. Advertisers, filmmakers, novelists, people who stood on stages. They made the story; the rest of the world sat on the receiving end and took it in. A brand told you who it was through a commercial or a slogan, and you believed it or you didn't, but the telling ran one direction.

That world is gone, and I do not think most businesses have fully absorbed how gone it is.

Today no one simply listens to a story. They step inside it. A brand is no longer defined by what it says about itself but by what a person feels while moving through it — and that feeling is assembled from hundreds of small moments, most of which no marketing department ever scripts. The people actually telling the story now are not the copywriters. They are the architect who decided where the light would fall and the front-desk clerk who decided whether to look up. They are the housekeeper, the person who answers the phone, the one who plates the food, the teenager pushing carts in the parking lot. Every one of them adds a sentence to the story, whether they know it or not. The barista's tone is a sentence. The uniform is a sentence. The response to a complaint is a whole paragraph. And the story is no longer told in a single moment; it accumulates, across every touchpoint, until it becomes something a person carries for the rest of their life or forgets by the time they reach the parking lot.

This is the shift that undoes the old model. You cannot control a story that is being told by five hundred hands. You can only align it — make sure that every one of those hands understands the same true thing about who you are, so that whichever touchpoint a person happens to encounter, they meet the same character. The new storytelling is not authored. It is orchestrated. And when it is orchestrated well, the effect on a human being is profound.

I want to show you what it looks like when an entire place tells one story, coherently, through every hand that touches it. And I want to tell it as what I was that week — a guest, yes, but a builder too, unable to stop watching how the thing was made even while it was working on me.

In the winter of 2020, my family traveled to The Ranch at Rock Creek in Montana. I have told you already how the place reached us before we arrived, through a website that built the anticipation for weeks. This is what happened when we got there — and it happened during the strangest season any of us had lived through, when the whole world had been instructed to stay apart, and gathering felt at once precious and slightly illicit.

The ranch sits on more than six thousand acres near Philipsburg — rolling meadows, towering pines, icy streams, mountain vistas that stretch until they dissolve. The setting alone does a great deal of the work: vast skies painted in winter pastels, horses grazing behind split-rail fences, a stillness so complete it makes you lower your voice without deciding to. It is the American West as we imagine it in novels and films, and somehow more so in person. A place like that could coast on scenery. Its whole genius is that it refuses to.

Because when we arrived, the ranch was nearly empty. So many guests had canceled that the expansive property had gone hushed and reflective, and I watched, with real professional curiosity, to see what the place would do with a problem that could easily have deflated the entire experience.

What it did was quietly astonishing. Rather than let the emptiness thin the experience, the staff concentrated it. Everything narrowed around the few of us who had come. Activities were curated for our family alone. Because of the precautions of that year, every outing was arranged privately, which meant that three generations of our family could be together, safely — those in their nineties included — in a way almost nothing else that year allowed. What could have felt like a diminished, half-staffed off-season instead felt like the ranch had been built for us. The constraint became intimacy. That is not luck. That is a brand that knows its story well enough to keep telling it when the script falls apart.

The house we stayed in taught me something before anyone said a word. A rustic, elegant ranch home with a wide front porch lined with rocking chairs facing west, so that at dusk the whole sky performed — amber, then lavender, then a deep and final blue. Inside, nothing was polished for effect, and that restraint was itself the design. Worn leather chairs softened by actual years. Thick woven blankets thrown over the sofas. Reclaimed beams gone dark with character, iron fixtures, fur throws, old saddles, weathered trunks, vintage photographs of people whose names no one remembered — every object chosen as though it had a history to tell.

Even the lodge bar carried the story: real saddles for barstools beneath heavy timber, chandeliers wrapped in antlers and evergreen. It was immersive and never once theatrical, which is the hardest line in this entire craft to walk. Theatrical says look what we made. Immersive says you belong here. The ranch never crossed the line, and I have built enough environments to know how much discipline that restraint requires. Theatrical is easy and cheap and everywhere; it announces its own effort and asks to be admired. Immersive hides the effort completely, which costs far more and is noticed far less, and is the only one of the two that ever makes a person feel at home.

Over several days I realized there was not a square foot of that property that had not been considered for how it would make a person feel. The textures, the scent, the light, the music, the furniture, the architecture — all of it pulling in one direction, telling one story. The ranch was not selling luxury in the ordinary sense of thread counts and price tags. It was selling immersion. Story. Belonging. And it was doing it not with grand gestures but with relentless, accumulated attention.

The service is where the story became unmistakable, and where I want to slow down, because the details are the whole lesson.

Our first evening, we followed softly lit paths to an outdoor spa set into the winter landscape — the shock of crisp Montana air against bubbling warm water, oversized robes, thick towels, glowing lanterns, snow dusting the trees around us. It was sensory storytelling at its most complete, and it asked nothing of us but to be present inside it.

Later, back at the cabin, we found slippers lined up on a bench, games waiting on the table, and a refrigerator stocked with things we had merely mentioned in passing earlier in the evening. No announcement. No flourish. Just attentiveness — the kind that makes a person feel not served but seen.

The ranch is known for its service, but what stayed with me was never efficiency. It was intentionality, and it ran on a single quiet philosophy: avoid saying no whenever possible. One evening my sister mentioned, offhandedly, that she was craving eggnog. The ranch did not have any. By the next afternoon, when we came in from an excursion, a fresh carton was sitting in our refrigerator. Another evening we discovered the television in our cabin could not take an HDMI cable we needed. In a remarkably short time, the television itself — not the cable, the television — had been replaced with one that could. That is not routine hospitality. Nobody escalated those requests to a committee. Someone heard, and someone acted, and the acting said what all real attentiveness says: your small wish registered as important to us, because you are important to us.

Notice that none of these moments were expensive in any way that shows up cleanly on a page. A carton of eggnog. A swapped television. A refrigerator stocked with the snacks we happened to name. What made them powerful was not their cost but their coherence — every one of them told the same story as the beams and the saddles and the sky off the porch. This is Montana. This is ranch life. This is your story now, and we are paying attention to how it feels.

And the days themselves carried the same narrative, seamlessly, one experience flowing into the next. We rode horseback through icy streams and over snow-covered trails. We bundled under blankets in a canvas-covered wagon along old mountain roads. We learned to shoot skeet, crossed the snow on snowmobiles, searched for animal tracks, and came in at night to roaring fires to thaw our fingers and tell the day back to one another. We bowled. We watched movies under warm blankets in the lodge. Nothing felt like a menu of amenities. Everything felt like chapters of the same book.

If the ranch shows how coherence works over days, a different place taught me how it can be conjured in a single, surprising stroke. Some years earlier, during a holiday visit to Austin, Texas, I saw the Four Seasons build what it called an Ice Rodeo — an outdoor skating rink, of all things, in the middle of a city far more associated with cowboy boots than with winter sports. Twinkling lights, seasonal music, festive cocktails, fire pits, and a hundred carefully layered details turned something that made no obvious sense into something people could not stop talking about. It worked precisely because it was a surprise that still cohered — unexpected, and yet completely of a piece with a brand that promises delight.

That is the other face of the new storytelling. Coherence does not mean predictability. The best storytellers can astonish you and still, in the same motion, feel exactly like themselves. Rock Creek did it by never breaking character across a week. The Four Seasons did it by breaking every expectation about Austin in December while never once breaking character about who they were.

I keep coming back to one thing these places understood that most businesses never quite do: people do not remember perfection nearly as much as they remember how a place made them feel. They were not chasing flawlessness. They were chasing feeling, and feeling is built from touchpoints that all agree with one another. When the environment, the staff, the activities, and the smallest gestures are all telling the same emotional story, something crosses over. The brand stops being observed and starts being felt. That is the line every builder is trying to reach and most never do — the moment a place stops presenting itself to you and simply surrounds you.

I felt it most on an evening I did not expect to. Three generations of my family, together in a warm room while the snow came down outside, in a year that had told all of us to stay apart — and here we were, held. My relatives in their nineties among us, all of us inside the same hour, the same fire, the same story. The ranch had made that possible not through any single dramatic act but through a thousand small decisions that added up to safety and warmth and belonging. And I sat there, the builder in me and the woman in me finally in the same chair, and understood the whole of my professional life from the guest's side of the counter. This is what I had spent twenty-five years trying to make for strangers. This is why the work mattered. Not because it moved product. Because it could take a frightened, scattered season and hand a family an evening they would keep forever.

That is the power of what I have come to call the new storytellers. They do not communicate a message. They build a world so coherent that a person forgets they are inside a designed thing at all — and simply feels, for a while, that they matter here. Every employee embodied the story. Every corner of the environment reinforced it. Every small gesture personalized it.

And the brand was not something we looked at. It was something we lived inside.

Six thousand acres is amazing but that's not the magic. You need to understand what these places understood: that you are always telling a story, through every hand and every detail, and that its power comes entirely from coherence — from making sure every touchpoint says the same true thing. A home can do this. A small shop can do this. A single evening for people you love can do this. The scale changes. The craft does not.

*The eggnog and the swapped television have stayed with me for years, and I finally understand why: in both, someone was empowered to say yes. That is not a small thing, and it is not an accident of good staffing. It is a design decision — one that most organizations, and most of us in our own homes, quietly refuse to make, because yes is harder and riskier than no. The next chapter is about building the architecture of yes: how you give the people closest to the moment the freedom to make it matter.*

CHAPTER 19

# The Architecture of Yes

*The most memorable moments often come down to a single thing: someone was empowered to say "yes."*

THE EGGNOG THAT APPEARED OVERNIGHT AND THE television that was swapped instead of the cable have stayed with me for years, and I have finally understood why. It was not the eggnog. It was that somewhere in that organization, a person heard a small wish and was free to grant it — without a form, without a manager, without a meeting. Somebody had built, into the very bones of the place, the freedom to say yes.

That freedom is rarer than it should be, and its rarity is not an accident. Most organizations are quietly, structurally built to say no.

I want to be fair about why, because the instinct is not villainous. No is safer. No is consistent. No protects the company from its own people — from their mistakes, their generosity, their judgment. A rule says the same thing to everyone, which feels like fairness and is certainly easier to manage. So businesses write policies, and the policies harden, and slowly the people closest to the customer lose the authority to do the obvious human thing standing right in front of them. They can see exactly what would make the moment right. They are simply not allowed to do it.

You have watched this happen, and it is one of the most quietly demoralizing experiences in modern life. Picture the gate agent at a delayed flight, or the person behind the returns counter, or the voice on the customer-service line. Something has gone wrong, and the fix is obvious, and you can see in their face that they know it too — and they cannot do it. "I'm sorry, there's nothing I can do." The tragedy of that sentence is that it is usually true. There really is nothing they can do, because someone above them decided that the risk of a human being using judgment was greater than the cost of a thousand people walking away feeling like a case number. That is the architecture of no, and what it communicates, every single time, is unmistakable: a system matters more here than you do. Both people in that exchange feel it. The customer feels unimportant, and the employee feels like a machine that has to apologize for its own programming. Nobody wins. The rule wins.

And we have found a way to make that sentence permanent. For a century, the architecture of no was at least staffed by human beings, which meant it could always, in a pinch, be overridden — a supervisor summoned, an exception granted by someone who could see your face. We are now automating the frontline, and an automated frontline cannot say yes. A chatbot has no discretion to extend. A phone tree has no judgment to exercise. An algorithm trained to minimize cost will not — cannot — decide that this particular person, tonight, deserves an exception. The danger of automating service is not that the machine will be rude. It is that the machine industrializes "I'm sorry, there's nothing I can do," and removes the one thing that used to redeem that sentence: the human being who occasionally broke the rule out of mercy. I will come back to this, because it is one of the central questions of the age. For now, simply notice this: the more of our touchpoints we hand to systems that cannot say yes, the more valuable the ones that still can.

The architecture of yes is the deliberate decision to build the other way. And beneath it is a single conviction I have staked a career on: you cannot script humanity. You can only build the conditions in which people are free to exercise it. Every practice in this chapter is a way of doing that one thing.

The most elegant example I know fits, more or less, on a single card. For years, Nordstrom's employee handbook was famously said to be one line: *Use best judgment in all situations. There will be no additional rules.* The truth is a touch more ordinary — the card always came alongside the usual legal paperwork, and in time a fuller handbook joined it — but the paperwork was never the point. The point was the sentence, and the culture it announced. Think about what that one line builds. It converts every employee from a rule-follower into a judgment-maker. It tells them, before any customer ever appears, that their read of the human being in front of them is trusted more than any policy could be. And it produces the legendary moments — not because a rule required them, but because a person was free to see and to act. The most famous of those moments is true, and I love that it is: in 1979, a man walked into a Nordstrom in Fairbanks, Alaska, to return two tires. Nordstrom does not sell tires and never has — but that store had taken over a location that once did, and rather than explain why the return was impossible, the salesman worked out what the tires were worth and simply handed the man his money. No rule required that. A rule would have forbidden it. A culture permitted it.

There is another way to build the very same thing, and Costco is the example I trust most, because I am one of its most loyal customers. Costco has baked its architecture of yes not into individual discretion but into policy itself. Its return policy is famously, almost recklessly generous — bring most things back, whenever, with no drama and no interrogation, and they will simply make it right. The generosity has a few sensible edges: electronics carry a ninety-day window, and flagrant abuse — the item hauled back after years of hard use, the membership built on serial returns — will get your card quietly revoked. But within those edges, the message is the same one Nordstrom's sentence sends, only structural: we would rather occasionally be taken advantage of than ever make an honest customer feel doubted. And here the obvious objection deserves an honest answer: yes, some people will try to abuse a generous yes — which is exactly why the generosity has edges rather than none. Costco has run the numbers and found that the cost of the occasional cheat is trivial next to the value of a customer who feels, bone-deep, that they will be treated fairly. Trust is not naïveté; it is a wager, with a few guardrails — and it is one of the best wagers in business. I will buy almost anything there, tires and toothpaste and a television, because I know that if it disappoints me, the company will stand behind it.

There is a third way still, and it may be my favorite, because it puts a number on the trust. At the Ritz-Carlton, employees — every one of them, from the front desk to housekeeping — are empowered to spend up to two thousand dollars, per guest, per incident, to solve a problem or make a moment right, without asking a manager. Two thousand dollars. No permission required. This is not a relic of some gentler era; it is still the policy today. And most of them never spend anything close to it, which is precisely the point. The point is what the number *says* to the person holding it: we trust your judgment with real money and real consequences, because you are the one standing in the moment, and we would rather you act than ask. A housekeeper who can send a child's forgotten stuffed animal home by overnight courier — with a letter written as if from the toy, describing its extended vacation at the hotel — is not following a script. She is using a freedom the company deliberately placed in her hands.

A culture of yes can live in a single empowered employee, in a number written into policy, or in a promise so generous it removes the need to ask. Every one of them communicates the identical thing to the person on the other end: we trust you, and we are on your side. That trust is never marketed into anyone. It is built, one no-questions yes at a time, until doing business with you simply feels safe.

That is the whole secret, and it is worth saying plainly, because it runs against every managerial instinct: you cannot script your way to a great experience. You can script your way to a consistent one — reliably fine, reliably forgettable — but the moments that actually move people, the ones they tell stories about for years, are almost never in the script. They are improvised, in real time, by someone who noticed something and was allowed to respond. These are not policies executing. These are humans, trusted, doing what humans do when you free them to care.

So the design question is not *how do I write a rule for every situation*. It is the opposite. *How do I build a place where the people closest to the moment can say yes?* And the answer has three parts, none of them expensive.

First, you hire for the thing you cannot train. You can teach almost any procedure. You cannot easily teach a person to notice — to sense that a guest is exhausted, that a customer is embarrassed, that a moment has quietly gone wrong. That attentiveness is either in someone or it isn't, and it matters more than any credential on the résumé. I have hired a great many people over the years, and the ones who made places come alive were almost never the most polished on paper. They were the ones who, in the interview, noticed something about *me* — who treated the exchange as human rather than transactional. Hire the people who see other people. Everything else is teachable.

Second, you give them real authority and you mean it. Latitude with a catch is not latitude; people can smell a permission that will punish them for using it. If you tell your team they can make it right, you have to let them make it right — and you have to let them occasionally make it right in a way you would not have chosen. Because the alternative, clawing the authority back the first time it costs you something, teaches everyone in the building to stop trying. Nothing travels through an organization faster than the story of the person who used the freedom they were given and got punished for it. One such story undoes a hundred speeches about empowerment. Trust that is withdrawn under pressure was never trust. It was a test, and people know the difference.

Third, you celebrate the yeses out loud. Every organization tells stories about itself, and those stories teach everyone what is actually valued, no matter what the mission statement claims. If the stories your people hear are about the time someone got in trouble for overstepping, they will learn to keep their heads down.

If the stories are about the time someone did something generous and unscripted and made a stranger's whole day, they will learn to look for their own version. Culture is not the values on the wall. Culture is which stories get retold in the break room. You get more of whatever you honor — so honor the yes. Tell those stories yourself, on purpose, until everyone knows what kind of moment earns applause here.

I learned this from the other side, as someone who spent years building teams to make places feel alive. The magic I most admired in my own career — the hand-built holiday worlds, the crews who started in July to have Christmas ready by December, castles and forests and animated bear orchestras made by our own people rather than ordered from a catalog — was never made by anyone following instructions. It was made by people who had been trusted with a vision and turned loose to realize it, who cared because they had been given something worth caring about and the room to do it their way. I watched what happened to that same kind of work, later, when the latitude was stripped out in the name of efficiency and control. It went flat, instantly and unmistakably. The castle became a kit. The magic became a line item. And the people who had once poured themselves into it learned, quickly, to stop. Nothing kills the yes faster than a manager who cannot bear the risk of it.

And here is where it comes home, because this is not only a lesson for businesses. The architecture of yes is something we build, or fail to build, in our own lives and relationships all the time. Think about how often, with the people closest to us, we default to no — not out of cruelty, but out of habit and self-protection and fatigue. The reflexive *not now.* The rule enforced because it is a rule. The small request refused because saying yes would cost us a little more effort or comfort than we felt like spending.

We build, without noticing, a private architecture of no, and the people we love learn to stop asking. A child stops bringing us the drawing. A partner stops suggesting the thing. The yeses we withhold are rarely dramatic. They are the extra ten minutes, the detour, the small want granted just because it mattered to someone we love. But they accumulate exactly the way the small yeses did — into a felt sense of whether, in this place, with this person, one's small wishes are safe.

I am not arguing for yes to everything; a life of no boundaries is its own kind of neglect, and a parent who cannot say no is not building anything either. I am arguing for noticing how often our no is reflexive rather than chosen, and how much of another person's sense of belonging is built from the moments we decided their small wish was worth granting. The architecture of yes, in a home as much as a hotel, is simply this: being the kind of presence where people are free to ask — and are sometimes, delightedly, met.

CHAPTER 20

# Know Thyself

*It is one thing to decide what your brand is; it is another to live it — every day, with every customer.*

I HAVE ALWAYS WANTED TO EXPERIENCE A EUROPEAN Christmas Market — an historic outdoor festival of small wooden stalls filled with the fragrance of holiday goodies, fresh pastries, decorations, trinkets, and souvenirs. While I had been to Europe on a couple of occasions, it had always been in the Summer or Fall. So heading into Canada for the holidays was a bucket list adventure.

In the Winter of 2018, I had the opportunity to fly north from Charlotte with my family to celebrate Christmas in historic Quebec City. Having worked in shopping centers for so many years, I was fairly burned out on Christmas — some years I didn't even put up a tree. But this year, the idea of visiting a Christmas market so similar to Europe's was intriguing. This trip, I had been so busy with projects, I left the planning up to my sister. I was just along for the ride.

And what a ride it was. Quebec City goes all out during the holidays. The famous Fairmont Le Château Frontenac outdoes itself with decorations that include a festive castle, lavish lobby decor, and programming that would put any Grinch in the spirit. I assumed that is where we would stay so I was surprised when my sister mentioned that she had found a boutique hotel nestled among cafés, restaurants, and art galleries.

It turned out the hotel my sister chose was Hotel Du Vieux, a 45-room elegant hotel with personalized service and a gracious atmosphere. The spacious rooms delivered with comfy beds, stone walls, large windows, and low lighting. Just the kind of place you want to huddle down in after a long flight. And that's just what we did.

To my surprise, the next morning I heard a rustling in the hallway. A large picnic basket was hung from my door. Opening it was like getting my first Christmas gift. The smell of fresh croissants and the pastry of the day along with a thermos of steaming hot coffee, orange juice, and an assortment of cheeses was just the present I needed to start my day. As I learned over the next few days, the hotel prides itself in exceptional service from the front desk to every staff person we met during our stay. This is a business that understands their customer. The concierge summoned carriage rides, made arrangements for a dog sledding and snowmobiling adventure, and seemed to know every restaurant that would appeal to weary but excited Americans.

As night fell, our group was told that there would be a church service at the small town perish. And, then we were advised to visit the Christmas Market for hot chocolate and homemade candies. I could have sworn I was standing in the square of Salzburg. Pink cheek children squealing with delight as they put sticky fingers of caramel into their mouths. Apron-clad gentleman telling stories and ringing bells seemed like they could be working as part of the Claus team. Horse-drawn buggies filled with visitors wrapped in blankets pulled up curbside. It was so incredibly cold. And, so incredibly beautiful.

The next day, those with adventurous hearts headed over to the 800-foot icy wooden toboggan slide. Our most senior family member announced that he would go first. That's usually me but not on this trip and certainly not in this degree of cold. Later, we walked the small alley streets draped with fresh greens and the sounds of shopkeepers chatting up the occasional tourist who wandered in.

It was an amazing trip. What stood out to me was how impressed I was with the little hotel. They understood how meaningful great service was to their guests. They knew that they didn't compete on size, amenities, or glamour. Their value was on making each guest feel like they were home. Home in a foreign country. Home for the holidays. Home with family. They did that one thing and they did it incredibly well.

*That European Christmas Market scratched one item off my bucket list. It was what I thought would make the biggest impression on me. As I thought about it later (and still talk about it year's later!), I realized that it was the hotel that really made the difference. I think the lesson for me is to know your strength and play to it. It is one thing for an agency to decide what your brand is — it is another to live it every day with every customer. Those are touchpoints layered on to create an experience like no other.*

CHAPTER 21

# Ritual and Return

*A wonderful moment is a memory. The same moment, repeated until it can be counted on, becomes belonging.*

A SINGLE WONDERFUL MOMENT IS A MEMORY. THE SAME wonderful moment, repeated until people can count on it, is something else entirely. It is a ritual — and ritual is the most powerful tool anyone has for turning a place, or a relationship, or a life, into something that holds. Everything I have described so far in these chapters — the details that speak, the recovered failure, the anticipation, the freedom to say yes — reaches its full power only when it repeats. Because belonging is not built in a moment. It is built in a rhythm.

I learned this most clearly at Birkdale Village, and since I have already told that story in full — the cow pasture we made into a village, the Fourth of July water fight, the traditions that outlasted us — I will not retell it here. I shared it, in that earlier chapter, as proof that belonging can be built on purpose. I want to return to it now for one narrower reason, because folded inside those traditions is something I did not stop to name at the time: why repetition works at all.

It is worth being exact about why this matters so much, because it is the mechanism underneath nearly everything in this book. Repetition does something no single event can. A first visit is an experience. A second is a coincidence. A third is the beginning of a pattern, and the human mind treats a pattern as a kind of promise — this will be here, this can be counted on, this is part of how my year is shaped. Something profound happens when an experience becomes predictable in the good sense: it stops being something you evaluate and becomes something you belong to. You do not stand outside your own traditions grading them. You are inside them. And that is the quiet magic of ritual — it moves an experience from the category of things we judge into the category of things that are simply, reassuringly ours. Ritual takes the anticipation we talked about and gives it a schedule. It takes belonging and makes it renewable. A single visit can be bought; a rhythm has to be earned, one honored return at a time — which is why a tradition, once it truly takes hold, keeps giving back long after anyone is spending anything to sustain it.

The scale can be enormous or it can be tiny, and the mechanism does not care. I think of the neighborhood coffee shop where the person behind the counter starts your order the moment you walk in the door. It is such a small thing — a drink begun three minutes early — but it is not really about the coffee, and both of you know it. It is a ritual of recognition. It says I know you, you belong to the rhythm of this place, your arrival is expected and welcome. That is why people will pass two identical chains to reach the one where they are known. The product is the same. The ritual is not. And the ritual is what they are actually buying, though they would probably just say they like the coffee.

This is not a marketing insight. It is one of the oldest truths about being human, dressed up in a shopping village. We have always used ritual to hold time still — the harvest festival, the Sabbath, the birthday, the anniversary, the same carol on the same cold night. Long before anyone had a brand to build, human beings understood that meaning does not come from novelty. It comes from return. The moment repeated on purpose, until it becomes a marker in the year we can find our way back to.

I knew this in my own life long before I could have explained it, and I knew it from the losing end. My childhood was fifty moves and constant leaving; I have written about how little continuity I had. And the one place that held me — Chautauqua — held me precisely because it repeated. The same bells across the water, every fifteen minutes, every summer, the Westminster chimes carrying the hour over the rooftops and out across the lake. The same brick paths. The same friendships resuming each year as though no time had passed. I did not have to rediscover who I was there, because the rhythm of the place remembered me. That is what ritual does at its deepest. It hands you back a self you can return to. The bells were the purest ritual I ever knew, and they taught me — long before I had language for it — that time does not only pass. It can be marked, and rung, and made to hold.

So when I built traditions for strangers at Birkdale Village, and when I built them, earlier and clumsier, for my own daughters — the birthday parties, the trips, the cross-country drive I dressed up as an adventure — I was reaching, without quite knowing it, for the thing I had most lacked. I was a woman who lived in perpetual motion, who found it nearly impossible to be still inside her own life. Ritual was the one form of presence I could reliably manage, because ritual does not ask you to be spontaneously present. It builds the presence in advance, on a schedule, and then simply asks you to show up. For someone like me — for anyone who struggles to be here — that is not a small mercy. It may be the most reliable path back into your own life that exists. You do not have to feel present. You only have to keep the ritual, and let the ritual return you.

Which is the part I most want to hand to you, because it costs nothing and it is available tonight.

It's not the village or the fire truck. The most powerful rituals are almost always the smallest and the most repeatable. The particular way you say goodnight. The Sunday breakfast that is always the same. The walk after dinner, the song in the car, the phone call every week whether or not there is news. What makes these sacred is not their content. It is their return — the fact that they come back, reliably, until the people you share them with can count on them, and counting on them becomes a form of being loved. A tradition is just a touchpoint you decided to repeat. And repeated touchpoints are how a family, or a friendship, or a single ordinary life becomes a place that holds.

Build the rhythm, and the belonging follows. It always has.

*Those Birkdale Village traditions have now outlasted their marketing budgets, their staff, and the better part of a generation — which is the proof of everything Part Four has argued: that what is built from real, repeated, human touchpoints does not fade when the budget does. But that permanence is not guaranteed. It has to be protected, because there is always a pressure pushing the other way — a colder, more efficient logic that looks at the hand-built thing and sees only cost. I watched that logic arrive once and kill the magic overnight. Understanding it — and understanding why the human, handmade touchpoint still wins in the end — is the work of the chapters ahead, as we turn from how to build experience to why, in an automated age, it may be the most valuable thing you can build at all.*

PART FIVE

# A More Human Future

CHAPTER 22

# Not Seen

*To be processed instead of met is its own quiet heartbreak. This is what it costs to not be seen.*

THERE WAS A TIME I WAS, AT EVERY SINGLE STEP, NOT seen — because I have spent this entire book describing what it means to build the feeling of being seen for other people, and I think you should also see, plainly, what its absence costs. This one is personal, and it is recent, and I am still a little raw about it. But I have turned it over enough times now to know what it is actually about, and it turns out to be about something much larger than one bad week with one airline.

Let me start with the tradition, because the tradition is the point.

Each of my grandchildren, when they reach a certain age, gets a week with Gimme. That is me — the name is a fair confession, since I am a grandmother who dotes without apology. A week with Gimme has become its own small institution in our family. The child flies in, alone, and gets my full attention for seven days: special activities, adventures, and a healthy dose of the kind of grown-up, coming-of-age counsel that lands differently from a grandmother than it ever could from a parent. It is a ritual, in exactly the sense I described a few chapters ago — a repeated, anticipated touchpoint that has become a form of belonging for every child in our family. It is, in other words, the whole book, lived out in an airport pickup.

This particular time, it was my thirteen-year-old granddaughter's turn, flying in from Austin to Charlotte. Arranging travel for a minor traveling alone has to be done by phone, and it carries an extra fee — a hundred and fifty dollars plus tax, each way — for the airline's escort service. I paid it gladly. I was not buying a seat. I was buying the assurance that my granddaughter would be seen, and watched over, and handed safely from one set of caring hands to the next. That is what the fee is for. That is what the fee promises.

I arrived at the airport absurdly early, the way you do when someone you love is in the air. I picked up my pass at the Special Services desk, went through security, and made my way to the gate. When the plane landed, I was told there had been a last-minute gate change, so I hurried to the new one, knowing my granddaughter would be among the last off the plane, escorted by a flight attendant. I had my ID out to show the attendant. I had my camera up and running, because I wanted to catch her face the moment she saw me.

The passengers came off. Then the pilots. Then the last few stragglers. My camera was rolling.

And then off came my granddaughter — thirteen years old, a large escort lanyard around her neck — entirely alone. No flight attendant. No one to check my ID, because there was no one there at all. The gate desk was empty. There was no one to hand her to me, because no one had been keeping her in the first place. Later I asked her, gently, whether anyone had said anything to her as she got off the plane. She told me yes. Someone had said, "Thank you for flying American Airlines."

That was it. That was the whole of the escort I had paid for and been promised. A thirteen-year-old girl — alone in a busy airport, in a city known for trafficking — sent off with a pleasantry.

I went back down to Special Services to report what had happened, because a safety protocol had plainly failed. An employee told me to follow her down the lobby, and I did, and we waited. Twenty minutes passed. Then my phone rang — two airline attendants, audibly panicked, asking whether I had my granddaughter, and in nearly the same breath blaming her for having gotten off the plane without them. When the woman from Special Services standing beside me heard that my granddaughter was safe with me, she simply turned and walked away, leaving me alone in the lobby. The attendants on the phone hung up the moment they learned the child was accounted for. No one asked if we were all right. No one acknowledged that anything had gone wrong. The problem, from their side, was not that a child had been lost. The problem was now solved, because the child had been found.

It was a rough start to what was meant to be a special week. And I wish I could tell you it was the end of it.

On the return, we arrived at Charlotte well ahead of time. The Special Services attendant informed me, to my surprise, that my granddaughter's father could not be the one to receive her at the other end — it had to be her mother, who was working out of town. I had set up the original escort myself. I had supplied the father's information. No one had ever told me there would be a problem. After some argument, the attendant revised the form. What I did not know was that when they checked my granddaughter in at the gate, they removed that form from her lanyard. Then there was another gate change, this time to a different terminal. We walked to the new gate. They called her name — and told me she could not board, because the form was required and the form was back at the original gate, in the terminal we had just left, and the plane was now boarding. No one offered to help. No one offered to retrieve it. So there I was, a woman in my seventies, running through a crowded airport to fetch a piece of paper that their own employee had taken off my granddaughter's neck.

When I finally made it back, gasping, the plane had not boarded after all. It was sitting at the gate with mechanical trouble, where it would sit for over an hour. And they had known. I had heard an attendant at the previous gate say they were already trying to find another aircraft. They had time. They simply did not use any of it to help.

I reported all of it to the airline afterward, but because a child's safety escort had failed, and someone should know. They did not respond to the safety concern. They sent my granddaughter a small credit toward a future flight. And then they sent me an automated customer-satisfaction survey.

So I answered the survey, and I told them who they were writing to.

I told them that my history with this airline runs deep, and reaches back to its darkest day. In the summer of 1994, a jet came down in a thunderstorm on its approach to Charlotte, up from Columbia — USAir Flight 1016. In those days it was USAir; it would become US Airways, and then merge so completely into American Airlines that many of us can no longer quite remember where one ended and the next began. My mother was a volunteer chaplain at that airport, and I drove her there so she could help. We were both brought into a room filling with families, and we were each assigned to one, to sit with them while the news came in over the live feeds — that there were a few survivors, and that many had not made it. The family I was given was waiting on a young woman who had just left her own wedding in Columbia and was flying through Charlotte on her way to her honeymoon. I sat with them that night and into the next day as they learned what had happened to her. And when I prepared to go back to my job, the airline itself asked me to stay on and keep helping the families. They put handfuls of cash in my hands to take people shopping for the necessities they had arrived without. I went with them to the crash site. I sat through the long hours of weeping and phone calls and arrangements. I went with one family to choose a casket.

So when I wrote to this airline about a failure of safety, I was not a stranger filling out a form. I was someone who had once stood inside their worst hour and held their passengers' grief in my own two hands. I am certified in crisis communications; I know exactly how intense those moments are, and I know what it means to show up inside them. They had trusted me, once, at the very bottom. And when I finally trusted them — with one of the people I love most in the world — they failed me. Not once. Twice. The first time for safety. The second time for a plain refusal to lift a finger to fix a mistake that was entirely their own.

For all of it, they refunded half of the escort fees.

I want to be careful here, because it would be easy to turn this into a simple story about a bad airline, and that is not the story, and it is not fair. The individual people I met that week were not monsters. I am quite sure the gate attendant had been told, in no uncertain terms, not to leave her desk, and I do not blame her for following the rule she was given. Most of them, I suspect, were trapped inside exactly the architecture of no I described earlier — bound by forms and policies and a system that had quietly stripped them of the authority to do the obvious human thing standing right in front of them. The failure was not a failure of individual character. It was a failure of design.

And there is a larger pattern underneath it, one that is only going to intensify, which is the real reason I am telling you this story at all. More and more, the work of seeing customers is being handed off — to automated systems, to customer-relationship software, to satisfaction surveys, to policies built to handle the ordinary case efficiently and at scale. And for the ordinary case, that is fine, even good; I am not nostalgic for the days of waiting on hold. But the entire value of a human being in the loop reveals itself precisely when the situation is not ordinary — when something extraordinary happens, exactly as it did that week. An extraordinary situation cannot be met by a survey. It cannot be met by a policy. It can only be met by a person, empowered to see what is actually happening and to respond to it. And the quiet catastrophe of that week was that at every extraordinary moment, the person had been removed from the loop — replaced by a form, restrained by a rule, or answered with an automated survey. The one thing no system can do is see you when the script runs out. That is the moment a human being is for.

I know it does not have to be this way, because I have been on the receiving end of the opposite, from a company no smaller or less busy than an airline.

Not long ago I had a problem with a purchase at Target — an ordinary retailer serving millions of people a day, exactly the kind of operation with every excuse in the world to answer me with a form. Instead, someone from their Services and Engagement team sent me an email, and then followed it up with a hand-written note. It was warm, and specific, and human, and it left me feeling not processed but met. That is the whole of the point. Target is not a boutique; it is scale itself. If a company that size can choose to sound like a human being, then the size of the operation was never the reason the airline could not. Automation is not destiny. It is a decision, made over and over, about whether a person will be there when the script runs out.

I keep thinking about how little it would have taken. Because there is a question — four words — that changes everything, even when the rules genuinely tie a person's hands.

"How can I help?" I have learned over a lifetime that those four words put a person into a problem-solving posture, and that their effect on the other human being is out of all proportion to their size. Often the honest answer is that they cannot help the way you want. It matters far less than you would think. The willingness itself communicates the very thing we are all starving for — that someone sees us, that someone is listening, that someone is on our side. "There is nothing I can do" and "how can I help?" can lead to the exact same outcome and land on the human heart as opposite experiences. One closes a door. The other, even when it cannot open the door you wanted, at least leaves you standing with a person instead of a wall.

This is a truth worth noting. Run back through every one of those interactions and you will find that not one of them was, in the end, about a form or a gate or a fee. Every single one was a touchpoint that said the same thing: you are a case to be closed, not a person to be met. In person, by phone, by email, across two airports and one very long week, at no point was I seen. And I felt it exactly the way I have spent this entire book insisting people feel it — as the precise opposite of belonging, as the confirmation that I was not worth the attention.

This is what I would most ask you to take from all of this, whether you run an airline or a business or a family. What a story that could have been. Picture the other version. A gate agent who was there, who reassured a nervous thirteen-year-old and walked her to her grandmother. A single person, on the phone, who said I am so sorry that frightened you both — how can I help make this right? A reply to my letter that said we know exactly who you are, and we are horrified, and thank you for telling us. Any one of those moments would have cost the airline almost nothing, and it would have protected — even deepened — a relationship of decades. It would have turned a loyal customer into an evangelist for life, someone who told the story of the airline you can trust with your grandchild at every dinner table for the rest of her days. That story was theirs to keep, at every single step. They let it go at every single step. And the reason they will never get it back is the reason that sits at the center of this whole book: because being seen is the one thing a business cannot fake, cannot automate, and cannot afford to hand to a machine — and its absence is a cost that never shows up on any report, and is larger than any of them.

I spent my whole life learning to build the thing I did not get that week. I know exactly what it is worth, because I have watched it turn strangers into family and empty ground into home. And I know exactly what its absence costs, because I have now, at last, been on the other side of the counter — the customer, the grandmother, the loyal one, standing in a lobby while the person who could have simply asked how can I help turned, instead, and walked away.

*I have shown you now what it feels like to be a customer no one chose to see — and I have pointed at why an organization ends up that way, one reasonable, efficient decision at a time. No one at that airline set out to fail a grandmother and a frightened child. They were following a logic that could see the cost of everything and the value of nothing that matters — the same logic that once looked at a hand-built winter castle and saw only an expense. That logic has a name and a natural habitat — and understanding it is the key to why the human touchpoint is not a soft luxury but the most valuable thing a business can protect.*

CHAPTER 23

# When Optimization Breaks the Magic

*There are two ways to lose the thing that made you loved: cut it, or cash it in. What cannot be measured is the first thing discarded.*

I TOLD YOU EARLIER IN THIS BOOK ABOUT THE DAY THE magic died. How the mall where I learned my whole craft — Eastland, with its hand-built winter castles and its animated bear orchestra and its crews who started work in July — came to be run by Glimcher Realty Trust, a real estate investment trust; and how, for the holiday season of 2000, the productions we had built by hand were swapped, in a single decision, for the same catalog package every other mall in the country was ordering from the same vendors. I stayed one more year. In November of 2001, as the same generic decor was being installed for a second identical Christmas, I had seen enough, and I left. I said then that the magic did not fade gradually. It was replaced. And I said something I want to return to now, because it is the hinge of this entire chapter: the people who made that decision did not do anything

VILLAINOUS. THEY DID THE ORDINARY, SENSIBLE, MODERN thing. They just did not understand what they were throwing away.

I have thought about that moment for twenty years, because it is the clearest example I know of a mistake that businesses, and people, make constantly — a mistake that feels like prudence right up until it quietly hollows out the thing you were trying to protect.

Here is the shape of it. Someone opens a spreadsheet. On that spreadsheet, the hand-built castle and the catalog kit look almost identical. Both are "holiday display." Both draw children. Both photograph well enough for the local paper. But one costs a great deal more — the crews, the months of labor, the custom fabrication — and the other can be delivered on a truck. On the spreadsheet, there is no column for the feeling. There is no line item for the busload of schoolchildren who will remember that castle for the rest of their lives, no cell that captures the difference between a place that made something and a place that ordered something. So the spreadsheet says what spreadsheets always say. It says: these are the same, and one is cheaper. Choose the cheaper one. And a reasonable person, looking only at the spreadsheet, does.

The trouble is that the spreadsheet is blind to the only thing that ever actually mattered. It can see the cost of the snow. It cannot see the snow.

This is the central problem of differentiation in a commoditized world, and I want to state it as plainly as I can, because it is the closest thing this book has to a piece of hard business advice. Everything a spreadsheet can easily measure, your competitors can also measure — and match. Price can be matched.

Features can be copied. Convenience can be replicated, usually within a season. If you compete only on the things that show up cleanly in a column, you have entered a race that has no finish line and no winner, only a slow grinding descent toward the lowest price and the thinnest margin, where every player is interchangeable and no one is loved. That race has a name in my field. We call it commoditization, and it is where brands go to disappear.

The only durable escape from that race is the thing the spreadsheet cannot see. The feeling. The coherence. The accumulated human touchpoints that make a place or a product or a company genuinely different in a way that cannot be photographed and copied by next quarter. A competitor can match your price by Friday. They cannot match, by Friday, twenty years of a community decorating its children's bicycles for the same Fourth of July. That is not a feature. It is a relationship, and relationships are the one asset that cannot be cloned, because they were built in time and time cannot be rushed.

This is why I have spent a career arguing, to anyone who would listen, that the human and designed and intentional touchpoint is not a soft, decorative luxury to be cut when times get hard. It is the moat. It is the single most defensible competitive position there is, precisely because it is the hardest to build and the hardest to steal. Anyone can lower a price overnight. No one can manufacture belonging overnight. And belonging, unlike price, does not evaporate the moment someone else undercuts you — it is the reason people stay when the cheaper option appears, the reason they drive past the substitute to reach the real thing, the reason a place can lose its marketing budget entirely and keep its customers anyway. I watched exactly that happen at Birkdale

Village, where events we built two decades ago still draw families who have never seen an advertisement for them. That is what earned belonging does. It compounds, quietly, off the books, into something no competitor can touch.

But here is the cruelty of it, and the reason the magic keeps dying in places like Eastland: the moat is invisible on the very document used to decide whether to keep it. The spreadsheet shows the cost of the human touchpoint with perfect clarity, every quarter, in red. It shows the value of it almost never, because the value is diffuse and delayed and lives in people's hearts rather than in this month's numbers. So there is a permanent, structural temptation to cut the thing that differentiates you, because the cutting produces an immediate, visible saving and the damage arrives later, slowly, disguised as a general softening that no one can quite trace back to its cause. The castle comes down. Nothing bad happens that afternoon. The numbers even improve for a while. And then, over years, the place becomes indistinguishable from everywhere else, and no one remembers deciding to make it so.

I am not against efficiency. Efficiency is how a business survives long enough to matter, and there are a thousand touchpoints in any operation that should be automated, streamlined, made frictionless, precisely so that human attention can be freed for the places it counts. The error is not efficiency. The error is applying the logic of efficiency to the touchpoints that were never supposed to be efficient — the ones whose entire value is that a human being took time and care that a machine or a catalog never would. Automating those does not make them cheaper. It makes them disappear, and takes your differentiation with them. The art of building a lasting brand, or a lasting anything, is knowing exactly which touchpoints to make frictionless and which to protect from the spreadsheet with your life.

And this, like everything in this book, is not only about business. We do the same calculation in our own lives, all the time, and we make the same blind mistake. We optimize away the handmade things — the letter for the text, the visit for the call, the meal cooked slowly for the one delivered fast — because on the spreadsheet of our attention they look inefficient, and something cheaper will do. And it will, for a while. Nothing bad happens that afternoon. But the handmade touchpoints were carrying the differentiation of our relationships too — the proof that this person is not interchangeable to us, that we spent something on them that we would not spend on just anyone. Cut enough of them, and our closest bonds begin the same slow slide toward the generic that Eastland took, and for the same reason: we could not see, on the spreadsheet, the value of the snow.

The places and the people who last are the ones who protect the snow. Who know that the expensive, inefficient, uncopyable human touchpoint is not the thing you cut to survive. It is the thing you were surviving for.

Eastland is the version of this story most people fear: the magic cut away to save a dollar, the handmade thing the spreadsheet could not see, swapped for something cheaper. But cutting is not the only way a brand loses what made it loved. There is a second way, subtler and probably more common, because while it is happening it does not feel like a loss at all. It feels like success. You do not cut the treasured thing. You cash it in. And I have watched it happen, over the past two years, to a brand I have loved my whole life.

Let me tell you who I am in this story, because the stake matters. I am a thrifter. I came to it first out of necessity — a woman who moved fifty times and furnished a great many homes on very little money — and I stayed in it, long after the necessity passed, out of something closer to joy. Because somewhere along the way I discovered that the merchandise was never really the point. The hunt was the point. The particular thrill of walking into a cluttered, overstuffed, faintly dusty store with no idea what you might find, and then finding it — the one-of-a-kind piece, the thing someone else discarded, waiting there to be seen and rescued and brought back to life in a home that would love it. I have pulled treasures out of those racks that made interior designers gasp. And I have given back, too: across all those moves, I have donated thousands of dollars of goods to the same stores, so that what I released could go find its next life the way my finds had found theirs. I was, in every sense, the ideal customer. Loyal as a buyer, generous as a donor, evangelical to anyone who would listen. That is who they lost.

Because over the past two years, Goodwill got fancy.

I say that as gently as I can, and I want to be fair, because Goodwill does real good in the world and I am not interested in pretending otherwise. But the change is unmistakable, and I am not imagining it — longtime shoppers all over the country are saying the same thing, and retail analysts have started saying it too. Somewhere, someone did the math. They realized that the best donated items — the designer labels, the genuine finds — were being sold for a pittance on the same racks as everything else, and that this was, in spreadsheet terms, money left on the table. So they began pulling those items out. The good stuff now gets routed to online auctions, or to a growing number of curated, boutique-style stores designed to look less like a charity shop and more like a nicely kept version of the mall. Prices across the board have climbed, sharply. The stores have multiplied — there are now around 3,400 of them across the United States and Canada, over a thousand more storefronts than Target operates in the entire country — often under a scatter of different names, until in some towns you seem to pass one every few miles. And the racks themselves have been tidied into something that looks, feels, and prices like ordinary retail.

I will give you a sense of it, from my own recent trips. Very worn vinyl shoes, marked six dollars and eighty cents. A small, stained lampshade for five. A single wineglass for three — which is very nearly what I would pay for a new one online. Anything with genuine value gets pulled and priced on its own: a framed print at a hundred and twenty-five dollars. And the furniture is where it turns almost absurd — a damaged particle-board table at forty-five dollars, a mirror at fifty, a bed at three hundred. These are not treasure-hunt prices. These are I-could-have-bought-it-new prices. And the moment that becomes true, the entire reason to be there evaporates.

Here is a fact that makes this stranger and more instructive than it first appears. Goodwill is not really one company. It is a federation of independent local nonprofits, each with its own board, its own leadership, its own pricing and its own choices, all operating under one shared name and one shared logo. Which means no single villain decided to do this. It happened in pieces, region by region, each local organization making its own reasonable-looking decision to capture more revenue. But the customer does not see the org chart. The customer sees the logo. And so the choices of some quietly redefine the feeling of all — a brand spending down, storefront by storefront, a reputation it took generations to build, without anyone at any single location feeling responsible for the whole.

And what exactly did they cash in? This is the part worth slowing down on, because it is the heart of the matter. They did not sell off their inventory; they sell that every day, that is the business. They sold the treasure hunt. They took the one thing that made the experience unlike anywhere else — the serendipity, the thrill of the dig, the feeling of rescuing something valuable that the world had thrown away, and getting it for almost nothing, and doing a little good in the bargain — and they traded it for a higher price per item. On the spreadsheet, it looks like a win, and in the short run it very likely is one; revenue is surely up, because people still need to clothe themselves and furnish a home. But the feeling is gone. And the feeling was the product. Nobody ever loved Goodwill for its efficient retail merchandising. They loved it for the hunt, and the hunt is exactly what got optimized away.

There is one more layer here that Eastland never had, and it is the reason this particular self-sabotage cuts so deep. Goodwill is a mission brand. Its entire standing in the world rests on trust — you donate your things because you believe the organization does good with them, and you shop with a small glow of virtue because buying there felt like participating in that good, cheaply and cheerfully. Trust is not a nice-to-have for a brand like that. Trust is the whole business. And trust is precisely what erodes when a charity starts to feel like a slick retail chain that has priced out the very people it exists to serve, all while asking you, at every register, to round up your total to support its work. The saddest sentence a brand can earn is the one I now hear all the time about Goodwill: people say they still shop there, because they have to, but they no longer like the company. That is a brand living on its logo while it spends down its soul. And for an organization whose name is, of all things, literally goodwill, that is a loss almost too on-the-nose to believe.

Here is my place in this, because it complicates my complaint in a way that matters to me. When I first walked into those stores, decades ago, it was out of need. I was furnishing homes after moves I could barely afford, and the low prices were not a charming bonus — they were the whole point, the thing that let me make a place for my children out of what other people had let go. Today I go as a hobbyist, hunting beautiful things I do not need. So when the prices climbed, what I lost was a pleasure. But there are people for whom those same stores were never a hobby — people for whom an affordable winter coat, or a five-dollar pair of work shoes, or a cheap set of dishes is the difference between getting by and not. And it is their loss, not mine, that I cannot stop thinking about. My heart breaks a little for the shopper who comes to that familiar sign out of real need, the way I once did, and finds it priced like the very mall they already cannot afford. Whatever the balance sheets say about programs and community impact, something has quietly been taken from the people the whole thing was supposed to be for. That is the cost that never appears in the revenue. And it is the one that should keep someone up at night.

You have watched some version of this happen, I am certain, to something you loved. The neighborhood restaurant that got popular and slowly stopped caring. The favorite product that was "improved" into blandness. The local shop, full of character, that was bought by someone who saw only the margins and scrubbed the character out in the name of growth. It is a specific kind of grief, watching a thing you loved forget why you loved it — and it almost always arrives dressed as success, with the revenue rising even as the reason recedes.

So here is the lesson, and it is the one I would tattoo on the wrist of anyone building anything. The greatest danger to a brand is rarely a competitor. It is far more often the brand itself, forgetting why it was loved and optimizing that very thing away. The measurable is seductive precisely because you can see it — revenue per item, square footage, the number of stores. The unmeasurable is invisible on the report right up until the moment it is gone — the feeling, the trust, the treasure hunt, the reason anyone chose you over the cheaper, closer, easier alternative in the first place. If you build anything — a business, a practice, a table people gather around — you must know what your treasure hunt is. You must be able to name the intangible reason people come to you and not the identical thing down the street. And then you must protect it with everything you have, especially in the seasons when you could make more money by quietly selling it off. Because you can. It will even work, for a while. That is exactly why it is so dangerous.

Eastland cut the thing that made it loved; Goodwill cashed it in; but both failures share one blindness — an inability to see the value of the intangible, human, hard-to-measure thing until it was already gone. For a century that blindness cost brands their magic at the margins. Now a technology is arriving that turns the same question into the defining one of our age: when machines can imitate the human touch itself, convincingly and at scale, what happens to the intangible thing that was always our real value — and will we finally learn to see it before we trade it away? That is where the last stretch of this book is headed.

CHAPTER 24

# Who Can Make Music?

*The same technology that can counterfeit the human touch can also hand it back to us — if we ask it to.*

I HAVE BEEN HONEST, THROUGHOUT THIS BOOK, ABOUT what artificial intelligence might take from us — and I want to be just as honest, now, about what it can give. Let me tell you about a season when it gave something back.

During a really rough patch in his life, my brother discovered a platform called Suno. It lets an ordinary person — someone who has never written a note of music in their life — put words to melody and voice, and make a song. He wrote an album's worth of them. Not to sell, not to perform. To say the things he could not otherwise say, and to hear them sung back to him. It was, in the plainest sense, therapeutic. Music became a place he could set the weight down.

I understood the pull of it at once, because I was handed a great deal in this life but not, it must be said, the family gene for music. I come from a deeply musical family; I do not sing, and you can thank me later. I have never wanted to sing karaoke, or sit down at a piano in front of anyone, or imagine I could write a song. And yet a tool that could let me create in the one place I had always been locked out of — that got my attention.

Before long, I found a use for it. I was producing an annual event honoring the top agents in a luxury real estate company — a convention-center affair with a whole vendor village, our theme that year built around making magic. I brought in top hats and stuffed rabbits, fog machines and balloons and an actual magician. And for the opening, I decided to write a song.

I wrote the lyrics. I set them to melody, added a voice, produced the music, and played it as the agents walked in to be celebrated after a brutally hard year. Was it hokey? Completely — I knew it while I was making it. But something happened in that room I did not expect. People asked for copies. They emailed afterward. They were moved, not because the song was good, but because someone had taken the time to give them one. They had their song, for their moment. I had added a layer to that night I could never have managed before — a soundtrack of appreciation, made by the one person in the building least qualified, on paper, to make it.

So let me ask the question that gives this chapter its name, because it turns out to be a bigger question than it sounds. *Who gets to make music?*

For almost all of human history, the honest answer was: the gifted, the trained, the few. Music was gated — gated by talent you were either born with or not, by years of lessons most people could not afford, by instruments and studios and a whole apprenticeship of skill that stood between an ordinary person and the song in their heart. Plenty of us have carried a feeling we could not set to melody, a thing we wanted to say in the one language we were never taught to speak. The gate kept us out, and we made our peace with standing on the wrong side of it.

That gate is, quite suddenly, gone. And this is worth sitting with, because it is the most hopeful thing I know how to say about these machines. At their best, they do not replace a human capacity. They *distribute* one. They take an ability that used to belong to a lucky few and hand it to anyone with something to express — which means they can hand people the power to make a touchpoint they could never have made before. A song is a touchpoint. It is one of the oldest we have. And a technology that lets a grieving man, or a woman who cannot sing, make one anyway is not stealing anything human. It is widening the circle of who gets to be human out loud.

Now — I will not pretend the story is simple, and I would not respect you if I tried. There is a serious controversy around tools like this, and it deserves to be named plainly. Artists and record labels have sued Suno, arguing it learned to make music by feeding on their work without permission — and the evidence that it trained on an enormous trove of copyrighted recordings appears to be substantial. Some of that fight has since resolved into deals: Warner Music settled with Suno in late 2025 and began moving toward licensing rather than litigation. Others have not. Universal and Sony are still in court, with major rulings expected through 2026, and the outcome will help decide the rules for this entire new art form. These are not small objections, and I hold them alongside everything else I am about to say. The people whose life's work trained these systems deserve to be seen, and credited, and paid — and I say that as someone whose entire book is a plea for people to be seen. A book like this one cannot look away from creators asking, essentially, to be seen. I hope the industry gets that part right.

But I want to draw a line that I think matters, because two different questions keep getting tangled together. *How the tool was built* is one question — a hard, unresolved, genuinely important one, and the courts are right to press it. *What the tool can do for a human being* is a different question. And on that second question, I can only tell you what I watched from the inside. I saw music return to my family's life after a long, long silence. I watched a hurting person find a way to speak. I watched a room full of exhausted people light up because someone had turned their year into a song. More than a hundred million people have now used Suno to make something — first-time creators and Grammy winners alike — and whatever the courts ultimately decide about the training data, that hunger to make and to be heard is real, and old, and human.

And yet I meet people almost every week who have decided they will never touch it at all. I understand the fear; it is real, and I will not wave it away — some of it will come true, and the people whose livelihoods are genuinely on the line deserve far more than a lecture about progress. I notice only, and gently, that most of us who swear we will never use AI are already leaning on it all day: the map that reroutes us around the wreck, the bank that catches the stolen card, the inbox that quietly eats the spam. That fear is worth taking seriously, and I mean to, before this book is done. For now, I will only tell you what this one small song taught me.

Here is the line I keep coming back to, and it is the truest thing I have learned about any of this. The question is never the tool. It is what we ask the tool to do. The very same model that can hand a grieving man his voice can also write the birthday message you did not mean, to the person you did not think about, in words that cost you nothing — a counterfeit of care, presence faked rather than given. The danger and the gift live inside the same machine. Which one we get is not up to the machine. It is up to us. A tool can fake a feeling, or it can help a real person express a true one, and nothing in the technology decides which. Only the human holding it does.

And that is the part that belongs in this book. The deepest fear I have carried through these pages is that the machines will make us less human — that they will counterfeit our presence and hollow us out. Here was the opposite. Here was a tool that did not replace a human capacity but returned one; that did not fake a feeling but helped a real person express a true one.

That is not the machine stealing our presence. That is the machine handing it back. The tool was never the point. The point is the man who could finally say what he felt, and the room that felt loved, and the woman who does not sing — who will still never sing — but who, for one strange and wonderful moment, made music anyway.

CHAPTER 25

# The Fickle Suitor

*There is a difference between a company that serves you and one that has decided it is merely a service.*

HERE'S HOW OUR RELATIONSHIP ENDED.

For years, my favorite credit card was a Barclays card tied to my airline AAdvantage account. It wasn't glamorous, but I loved it, because every dollar I spent turned quietly into miles, and those miles turned into seats on planes, and those seats carried me to my daughters. I tracked those points the way some people track the weather before a trip — with anticipation. Of all the cards in my wallet, this was the one I reached for. We had, if a person can have such a thing with a bank, a good relationship.

And then I got dumped. Not just me, as it turned out. In December of 2024, American Airlines and Citi announced that Citi would become the exclusive issuer of the AAdvantage card, and that the Barclays version — mine — would be folded into the giant. In the spring of 2026, a notice arrived: my account was being transferred. It didn't say much. I waited for my new card.

It didn't come.

Here is what I didn't fully appreciate until I was living inside it: these were two very different creatures. I was being moved from a relatively small operation to one of the largest financial institutions on earth. I have excellent credit and I spend responsibly, and I confess I assumed I was exactly the kind of customer a bank would be glad to win — that there might even be some small gesture of welcome, some acknowledgment of a good account faithfully kept. Instead I had acquired a fickle and inattentive suitor.

Weeks passed. No card. I began to worry someone had intercepted it while I was traveling. I kept logging in to my old Barclays account for reassurance until, one day, without warning, it simply closed the door in my face and told me to go find Citi.

So I called. And waited. And called again. The system offered me a choice: hold for hours, or receive a callback within twenty-four hours. I chose the callback. It never came. I tried again. I made my way through a maze of chatbots that redirected me, misunderstood me, and more than once simply hung up on me. And then, at last, a human voice — and I want you to know I nearly wept with relief at the sound of it. But the voice was thick with an accent I couldn't parse, over background noise so loud I couldn't make out a single word. She transferred me. The line went dead.

I began again. Punch three. Punch two. State your issue. Hold. Another voice, clearer this time. Why are you calling? I need to know whether you've mailed my card. No. You'll get it in a few weeks. Click.

We have talked, in these pages, about the upside of these new tools, and I mean it when I say I'm grateful for them when they work. But there is a downside, and I had just lived it. Because sometimes you don't need efficiency. You need a human. Sometimes to fix the problem. Sometimes only to listen. Sometimes just to know where on earth to send you.

And this still astonishes me. To make the transfer, I had to close a fifteen-year account — fifteen years of credit history and a strong limit, the kind of relationship a person builds slowly and doesn't part with lightly. There was, I was told, nothing they could do about it. Nothing. A company that vast, and there was nothing they could do. I sat with the phone in my hand in a kind of quiet culture shock. This was simply a different way of being treated — a different way of doing business.

Change happens; I understand that. But brands should be aware, because there are always other options, and customers feel the difference between a company that is serving them and one that has decided it is merely a service. Those are not the same thing. You didn't just move my account. You changed my relationship with my money, my miles, and the way I'll travel to see the people I love. You didn't just lose a customer.

You lost a relationship.

CHAPTER 26

# Priceless Care

*A brand is never more than the people who carry it — and never more tested than at the edge of a life.*

SEVERAL YEARS AGO I WAS STANDING IN A BOARDING LINE, about to fly somewhere tropical for work, when my phone rang. It was my sister. Our mother had suffered another heart attack, and she was being taken to Cleveland Clinic.

I stepped out of the line and bought the next ticket to Ohio.

By the time I reached her she had already been treated and moved to a room — smiling, safe, and utterly at peace in a place she trusted completely. She would end up there more than once over the years. On one visit, having befriended the wife of someone in the hospital's leadership, she found herself installed in the VIP wing, in a beautiful room, treated like a queen with nothing to her name but a hospital gown and a small suitcase. You would have had to know my mother to appreciate the rightness of that. She had a servant's heart, and it shone through everything, undimmed by all the damage that heart had taken. She fit right in

— because the tenderness I saw in her, I saw in every employee, on every floor.

I spent those days at her bedside, and what I remember, almost as vividly as my worry, is my amazement at the people caring for her. The culture was something I had never felt inside an institution that size. Plenty of hospitals advertise "patient-centered care." Cleveland Clinic lives by a philosophy it calls Patients First, and you can feel it in the hallways. In 2007 it named the nation's first Chief Experience Officer — a role that did not exist anywhere in American healthcare until the Clinic invented it. Not to chase ratings, and not to gloss the place with hospitality, but because it had decided that healing means tending the whole person: the physical, yes, but also the emotional, the educational, the spiritual weight that patients and their families carry through the doors with them. That is not a marketing line. That is what actual healing requires.

The roots of it go back a century. When four physicians founded Cleveland Clinic in 1921, they built it on an idea radical for its time: that no single doctor could know everything, and that medicine should therefore be practiced as a group — physicians collaborating rather than competing, all of them arranged around the one person who mattered, the patient. A hundred years later that idea has grown into something staggering: some eighty-three thousand caregivers, sixty-six hundred physicians and researchers, nearly twenty-two thousand nurses and advanced-practice providers, across a hundred and forty specialties. And somehow, at that scale, the whole apparatus still bends toward a single human being in a single bed. One patient at a time, each one the primary concern.

My mother felt that. I felt it. People who cross the globe to be treated there feel it the moment they walk in.

I want to stay with that scale for a moment, because it is the same problem every institution in this book has faced, only here the stakes are absolute. How does an organization of eighty-three thousand people make one frightened woman in one bed feel like the only patient in the building? The honest answer is the answer this whole book has been circling: it does not happen by accident, and it does not happen through a slogan. It is built — deliberately, relentlessly — into the culture, and then delivered one caregiver at a time. A Chief Experience Officer, a founding philosophy, a hundred years of "Patients First" are the institutional scaffolding. But the promise is kept or broken in ten thousand small rooms at once, by the nurse who appears at three in the morning, the aide who adjusts a pillow, the doctor who sits down instead of standing over you. Every one of them, in every small moment, is either telling my mother *you matter here* or telling her she is a chart. There is no neutral, not even here. Especially not here.

And that is what gives this chapter its title, because in every other business in this book a mishandled touchpoint costs a sale, or a review, or a customer who quietly drifts to a competitor. Here it costs something you cannot price. A cold word to a person who is afraid does not lose a transaction; it strips a little dignity from someone who has almost none left to spare. A moment of real tenderness does not earn a repeat visit; it can be the thing a family holds onto for the rest of their lives, the memory that the worst day came wrapped in kindness. When the stakes are a life, the touchpoint is no longer a matter of marketing. It is a matter of ethics. The care is not the packaging around the medicine. In a way that is hard to say and truer than we admit, the care *is* medicine. I watched it at my mother's bedside: a patient who feels safe and seen is a patient more willing to trust, to ask the frightened question, to follow the plan, to hope — and hope, at the edge of a life, is not nothing.

And I have to be honest about something the tenderness of that hospital could not quite hide from me, because it is happening to medicine nearly everywhere, and it began long before anyone worried about artificial intelligence.

Somewhere in the last few decades, we lost the family doctor. Not all at once, and not through anyone's malice — through consolidation, and managed care, and the steady arithmetic of a system asked to see more people in less time for less money. The independent physician who knew your whole history, who had treated your mother and might one day treat your children, who could read you across a room because he had known you for thirty years, has been folded into ever-larger groups and networks, until "my doctor" quietly became "my provider" — often a different one each visit, meeting you for eleven minutes with your chart open on a screen between you. Much of this happened for real and defensible reasons: coordination, safety, access, the impossible economics of modern care. I am not pretending the old way was perfect, or that bigness is a sin. But something was traded away, and it was the oldest and most valuable touchpoint medicine ever had — the one where someone who knew you looked at you and said, *I've got you.*

You can feel the loss in a single small, modern moment. Results that a doctor once delivered gently, by phone, with a human voice ready to explain and reassure, now arrive on an app. Your phone buzzes on an ordinary afternoon, and there, between a coupon and a text from your sister, is a number, or a phrase, or a single word like *mass* — with no one attached to it. No voice. No context. No hand. The result is faster, and the speed is genuinely good, and sometimes it saves a life. But the touchpoint has been stripped of the human being who once made it survivable, and a frightened person is left to read the most important sentence of their year off a chart application, alone.

This is the depersonalization the whole book has been circling, arrived in the one place we can least afford it — and here is why I raise it now, at the end. Every force I have just named predates AI. And every one of them is about to be poured full of it. The same technology can go either way, and which way it goes is the entire question. Point it at the paperwork and the screen, and it can hand the doctor back the minutes the system stole — an ambient scribe that listens so the physician can finally look up from the keyboard and see you. That is a machine restoring presence. Point it at the person instead, and it becomes one more layer between you and someone who knows your name: the portal that never calls, the chatbot that triages your fear, the tireless and unaccompanied delivery of news no one should receive alone. That is a machine removing the last of it. Because a machine can read the scan — often better than the human eye. It cannot sit on the edge of the bed. It can calculate the odds. It cannot hold your hand while you hear them.

So when people ask whether we want our doctors using AI, I have come to think it is not quite the right question. The tool is never the point. The real question is whether we will use it to give our doctors back to us, or to finish taking them away. And Cleveland Clinic — for all that it is exactly the kind of enormous system we are right to be wary of — is the standing proof that the better answer is possible. Scale is not the enemy. Technology is not the enemy. The enemy is what both of them do when no one is left guarding the human relationship at the center. The Clinic guards it, on purpose, at the scale of a small city. That, in the end, is the whole difference — and it is a difference we will have to choose, deliberately, in every hospital and every tool and every frightened afternoon to come.

Because healthcare is different from every other business in these pages. The stakes are not loyalty or repeat visits or a beloved brand. The stakes are life, and sometimes death. And here is what has stayed with me: an organization that enormous holds its entire reputation in the hands of tens of thousands of ordinary people, each of them, in each small moment, either keeping the promise or breaking it. Their brand is not a logo. It is a promise of hope, and beneath it, a promise of care. And that — care delivered person to person, at the very edge of a life — is about as human as anything gets. My mother is the reason I know it. She arrived at that place with nothing but a hospital gown and a servant's heart, and a hospital the size of a small city made her feel, until the end, like the only patient it had.

CHAPTER 27

# A More Human Future

*When machines can imitate the human touch, real presence becomes the scarcest thing in the world — and the most valuable.*

FOR MOST OF THE LAST CENTURY, THE SPREADSHEET could only trim at the edges of the human. It could replace the hand-built castle with a catalog kit, standardize the greeting, thin the staff. But there were things it simply could not counterfeit — the intuition of a person who senses you are tired and slows down, the warmth that cannot be scripted, the sense of being genuinely known. Those remained stubbornly, expensively human, and so they remained safe. That is the line that is now dissolving. Because the force arriving in our lifetime can imitate not just the cheap touchpoints but the human ones, and it can do it convincingly, and it can do it at a scale nothing before it could approach.

I have tried, all through this book, to keep artificial intelligence as a pressure rather than a subject — a thing we look through rather than at. I am not a technologist, and the world does not need another confident prediction about where all this is heading.

What I can offer is a pattern, and a question that the pattern makes urgent.

Here is the pattern. I said at the very beginning that we are a species that has been handed extraordinary gifts before, and has not used them well. A century and a half ago the machines gave us back half our working hours, and Keynes looked at that trend and predicted his grandchildren would work fifteen hours a week and struggle mainly with the problem of leisure. The productivity arrived exactly as he foresaw. The leisure did not. We converted the freedom back into work, and into the buying of things. That was the first gift, and we largely declined it. Then, in a single generation, the internet made knowledge and connection instant and free — the second gift — and we spent a remarkable share of it on engineered distraction, on systems built to capture the very attention we were trying to protect.

And now the third gift is arriving, and it is the largest of the three, because it offers to hand back knowledge and time at once. Tasks that consumed careers can happen in seconds. Expertise that took decades to acquire can be summoned in a sentence. If the pattern of the last century holds, an enormous quantity of human hours and human attention is about to be freed. Which brings the question this whole book has been circling, the only question I think actually matters about the technology: we have squandered a gift of this kind twice now. What would it take to collect it this time?

Here's the counter-argument and it deserves more respect than it usually gets. Perhaps we did not squander anything. Perhaps, offered the choice between more hours and more comforts, we simply wanted the comforts more, and chose accordingly, and that was our right. Maybe the fifteen-hour week was never what we actually desired. I think there is real truth in that, and I do not stand in judgment of the choice. But I would ask this: now that we have seen how the story turns out — now that we know what it feels like to be handed abundance and end up more harried, more scattered, more absent from our own lives than the generation before us — would we choose the same way again? The gift is being offered a third time, and larger. The one thing we get, that our grandparents did not, is the knowledge of how we spent the last two.

Because here is what makes this gift different, and more dangerous, than the ones before. The first two liberations freed our time and our access. This one reaches for something more intimate. It reaches for the appearance of care itself.

I wrote earlier in this book about the difference between being recognized and being known — about how a system can learn our habits without ever knowing us, how a hotel app that remembers your room temperature is useful but is not the same as an employee who senses your exhaustion and quietly slows the whole interaction down. When I wrote that, the counterfeit of being seen was still fairly crude. It is not going to stay crude. We are about to live among systems that perform attentiveness flawlessly — that remember every preference, mirror our language, anticipate our wishes, and respond with what looks and sounds exactly like warmth. They will be tireless and frictionless and, in their way, genuinely impressive. And they will still not know us, because knowing is not the same as predicting, and care is not the same as its performance.

The danger, I have come to believe, is not that the simulation will fail. The danger is that it will succeed — that it will become so smooth and so convincing that we stop reaching for the real thing altogether. That we grow accustomed to being recognized and slowly forget what it felt like to be known. That we accept the performance of care because it is available, and easy, and never has a bad day, and asks nothing of us — while the real thing, the human thing, remains difficult and inconvenient and occasionally disappointing, the way real people always are. When the counterfeit is crude, we can tell the difference and we choose the real. The test comes when the counterfeit is perfect. That is the moment we are walking into. And the whole question of whether we build a more human future or a less human one comes down to what we do when the fake becomes easier than the true.

I know that for many of you, everything I have just described does not arrive as an argument. It arrives as a fear. I meet it constantly now — a flat, almost bodily refusal, *I will never use this*, said the way you might refuse something unclean. I do not want to talk anyone out of that feeling, because underneath it is something real and worth honoring: the fear of being made obsolete, at work and in some deeper way as a maker and a mind; the fear that the counterfeit will get so good the real stops mattering. Some of that will come true. The people whose livelihoods are genuinely in its path deserve a great deal more than a cheerful lecture about progress, and I will not offer one.

But I notice two things, gently, and I include myself in both. The first is that almost everyone who tells me they will never touch these tools is already leaning on them all day — in the map that reroutes around the wreck, the bank that catches the stolen card, the camera that rescues the photo, the search that finishes the thought, the inbox that quietly eats the spam. We draw our lines in strange places and defend them as principle. There is even a name for the discomfort of holding two contradictory things at once —

cognitive dissonance — and when it comes to technology, we are all a little fluent in it. The line most of us imagine we are drawing, between using this and refusing it, is really a line between the machines we have grown used to and the ones still new enough to frighten us. That line has moved with every generation, and it will move again.

The second is that a blanket refusal does not actually keep the technology out of your life. The world is going to be remade around it whether any one of us opts in or not. What refusal keeps out is not the tool. It is the good uses of it — the grieving man's songs, the doctor handed back her attention, the small moment of real help. To say *never* is not to protect your humanity. It is only to decline the chance to point the thing back toward it. And where we point it turns out to be the one question that was always, only, ours to answer.

Which is why I did not title this chapter with dread, and do not feel it — because we have stood in a smaller version of this doorway many times, and it has almost never gone the way the fear predicted. We did not stop going to concerts when the record let us hear the song at home; the recording made the live performance more precious, not less. We did not stop standing before paintings when we could see them on a screen; the print made the original, in its actual room, hit harder. We did not lose the joy of walking when we built the car and the bicycle; we learned that a hike is a thing you choose, on purpose, because the walking is the point. Again and again, a new way to reproduce or assist a human thing has not killed the real version. It has made it rarer, and so more sought, more valued, more loved. That is the deepest reason I feel hope and not dread: the very same force that makes the counterfeit convincing does something else at the very same time. It makes the real thing scarcer — and scarcity is where value lives.

Think it through. When knowledge is free, knowing things stops being valuable — what becomes valuable is wisdom, judgment, the human ability to decide what matters. When any task can be automated, doing tasks stops being the point — what becomes precious is everything a machine cannot do: to notice, to empathize, to create beauty, to offer real hospitality, to build trust, to help another person feel fully present inside a moment.

And beneath even those there is a floor the machine cannot reach, no matter how flawless the performance. A machine can generate an experience. It cannot live one. It can compose the condolence, but it has lost nothing. It can write the vow, but it will not keep it. It can say *I understand*, but it has never once been afraid. Everything that gives a human touchpoint its weight — that it cost the giver something, that it came from someone who shares your mortality, who remembers your history, who can be held to account tomorrow for what they promised you today, who could have withheld the care and chose instead to give it — all of that is precisely what cannot be manufactured. The machine offers the shape of presence with none of the stakes. And the stakes, it turns out, were always the point. We do not treasure being seen because the seeing is skillful. We treasure it because a person who did not have to, did.

These are not soft, sentimental leftovers. In an automated world they may become the most valuable capacities we have, precisely because they are the ones that cannot be manufactured. The more perfectly the machine can imitate care, the more a person will treasure the moment they encounter the genuine article — the touchpoint that is real, that cost a human being something, that could only have come from another heart. Genuine human presence is about to become the scarcest thing in the world.

This is the same truth I learned as a child, sitting inside the sound of the bells at Chautauqua, though it took me a lifetime and a desert to understand it. The bells did not tell me the time the way a phone does — extracting my attention, fragmenting my hours, demanding I check and check again. They gave the time to me, musically, across the water, so that I could receive it without ever leaving the moment I was in. That is the difference, in a single childhood memory, between a technology that steals presence and one that returns it. And it is the choice in front of all of us now, scaled up to the largest gift we have ever been handed. We can build a future of systems that monitor us, predict us, and quietly extract our attention while performing a care they do not feel. Or we can use the freedom the machines hand back to do more of the one thing they never will — to actually be present with one another. The tools do not decide that. We do. Every environment we build, every touchpoint we choose to keep human, is a vote for the kind of future we will live inside.

We have been handed this kind of gift twice, and let it slip. The question is whether, knowing that, we can finally choose to collect it — not the comforts this time, but the presence. It is the same choice, in the end, that the whole book has been asking, only now the stakes are the largest they have ever been.

CHAPTER 28

# Every Touchpoint

*Every moment leaves a mark. Some become part of who we are. Some become part of what we leave behind.*

AT THE VERY BEGINNING OF THIS BOOK, I SET TWO IMAGES side by side and told you it would live in the distance between them.

I have stood on a stage in front of ten thousand people, with jets crossing overhead and an orchestra swelling behind me, having built the entire spectacle myself. And I have stood in a grocery line, in a different season of the same life, without quite enough to cover the food in my arms. I said then that I had spent a career on the first — the wonder we can engineer for others — and that I was still, at this age, learning the second: the presence we so often cannot find in our own lives. I want to close the distance between those two images now, because everything in between has been the story of how they are actually the same story.

For most of my life I did not understand that. I thought the stage and the grocery line were opposites — one my triumph, the other my failure, evidence of a woman who could manufacture magic for strangers and could not manage to be steady in her own life. I built presence for other people with everything I had, precisely because I could not find it for myself. I was a woman who lived three steps ahead of every moment, scanning the road for what could go wrong, because that is what you learn to do when you have spent years in survival mode, when the present is where the danger lives. I could fill a town square with belonging. I could not, for the longest time, simply stand in my own kitchen and be there.

And then a car died in the desert, and there was no plan left to make, and for one night I was handed the thing I had needed to come to terms with my whole adult life — being present. I was fully present. Not because I had finally mastered it, but because everything that let me escape the moment had been stripped away, and there was nothing to do but stand inside my own life, terrified and alive, under an enormous sky. That night did not cure me. I have told you that honestly. I am still, to this day, a planner who has to be reminded — sometimes daily — to come back into the room I am actually standing in. But it showed me, once, what I had been chasing on all those stages. It showed me that the presence I could build for ten thousand strangers was available, even to me, even in a grocery line, if I could only learn to stop running long enough to receive it.

That is what every touchpoint in this book has finally been about.

What I most want to leave you with is the sentence the whole title has been reaching toward. Presence is not a mood that descends on the lucky. It is not a gift you either have or lack. It is built — the same way a brand is built, the same way a beloved place is built, the same way belonging is built — through touchpoints. Through the small, deliberate, repeated points of contact by which meaning is layered into ordinary time until that time becomes unforgettable. I spent a career learning how to do that for businesses and towns and strangers, and it took me until now to understand that the exact same craft is available for a life. Your life. The details that speak. The anticipation you build. The rituals you keep. The moments you refuse to automate away. These are not just how you make a place people love. They are how you make a life you are actually present for.

And it turns out you don't need a stage. That was the last thing I learned, and the most freeing. The whole apparatus — the jets, the orchestra, the ten thousand people — none of it was ever the point. The point was always the smallest thing: whether you are here, now, with whoever is in front of you. The presence I chased across a career of spectacle was waiting the whole time in the grocery line, in the ordinary unglamorous minutes that make up nearly all of any life. It was never on the stage. It was in the checkout line, if I could only have been there for it.

So let me say plainly what I believe, having spent a lifetime on both sides of this — building what I hoped would create presence and a sense of place for others and hungering for the real thing myself. My motto of "Being ready is not nearly as important as being real" came out of a very deep wound — one that many can relate to. I lived most of my years on the other side of that motto. I was so ready, so far ahead of every moment, that the moments passed without ever quite reaching me. The bells of my childhood tried to teach me the truth of it — that time does not only pass, that it can be marked and heard and made to hold, that the hour can come to you across the water if you are still enough to receive it. It took me most of a life to hear them. But they were right. Time does not have to be lost. It can be rung.

I am not going to pretend I have arrived. That would be the one dishonest note I could end on, and this book has tried, above everything, to be honest. I am still learning this. I still catch myself three steps ahead, still have to be called back into the room. What I can tell you is that the direction is right, and that the practice is available to anyone, at any age, in any life — that presence can be recovered, and built, one touchpoint at a time, by all of us, together.

So this is not a lesson from someone who mastered it. It is a summons from someone still walking toward it, and it is the same summons I keep issuing to myself. Build the moments that make your one life reach you. Notice the details. Author the anticipation. Keep the rituals. Protect the handmade things from the spreadsheet and the machine. Be here, in the checkout line and at the dinner table and on the ordinary Tuesday, for the people in front of you, while they are still in front of you. Every touchpoint is a chance to layer a little meaning into passing time. That is the whole of it. That is how a life becomes unforgettable, and how we become present for our own.

I am still learning this. I suspect I will be learning it for the rest of my life. I would be honored to keep learning it alongside you.

*This book began with a hunger — the human ache to be moved, to be reached, to be fully present in a life that so often slips past us unnoticed. It ends where that hunger has quietly been leading the whole time: not to a technique, and not to a stage, but to a choice available in the most ordinary moment of the most ordinary day. The touchpoints are always there, waiting to be authored. The only question the book has ever really asked is the one now in your hands, in your home, in your work, tonight and tomorrow and the day after: in a world engineered to keep us elsewhere, will we choose to be here? I am still choosing it, imperfectly, one touchpoint at a time. Come choose it with me.*

CODA

# A Place to Begin

*Presence is not a grand undertaking. It is something small you could begin tomorrow – if you are still enough to hear the bells.*

A BOOK ABOUT PRESENCE SHOULD NOT END BY ASKING you to do a great deal. That would rather miss the point. So I want to leave you with something small — a place to begin that you could start tomorrow, in whatever life you actually have, without a budget or a title or a stage.

Everything in these pages comes down to four movements, and they work the same whether you are building a town square or tending a marriage. You learn to recognize the touchpoints — to see that experience is always being authored, that every detail is already speaking.

You come to value them honestly — to understand that trust is built in the small things and broken in them too, and that attention is the whole currency. You build them on purpose — through anticipation, through coherence, through the freedom to say yes, through rituals that return.

And you leverage them, in the truest sense, by protecting the handmade human ones from everything that would trade them away for something cheaper and emptier.

Recognize. Value. Build. Protect.

That is the whole method, and you already have everything you need to practice it.

Here is how I would start, if I were you, this week.

Take the stranger's walk through one thing you care about. Your front door. Your team's first email to a new customer. The way dinner begins at your table. Come at it as someone encountering it for the first time, and ask the only question that matters: what is this already saying? Not what do I wish it said — what is it saying right now, to the person on the receiving end. Most of the time you will find it is saying something you never chose. That is not a failure. That is the opening.

Then change one sentence. Pick a single touchpoint — just one — and decide, on purpose, what you want it to say, and make it say that. Warm the welcome. Answer the message as though the person mattered, because they do. Set the table the night before. Begin the order before they reach the counter. You are not remodeling your whole life. You are authoring one detail that was authoring itself without you.

And then, the hardest and most important part: repeat it, until it becomes a rhythm someone can count on. A single good touchpoint is a nice moment. The same one, returning, is how belonging is built — in a business, in a family, in the relationship you have with your own passing days.

I cannot promise this will make you present all at once. It has not, entirely, for me, and I have been at it a long time. But I can promise this: every touchpoint you choose to author, and to keep human, is a small refusal to let your one life slip past unnoticed. Enough of them, layered into ordinary time, and the time stops slipping. It starts to hold.

That is where I would begin. Not with the whole of it. With one detail, chosen on purpose, tonight.

The bells are still ringing. You only have to be still enough to hear them.

*Every moment leaves a mark.*
*Some become part of who we are.*
*Some become part of what we leave behind.*

APPENDIX

# A Field Guide in Brief

EVERYTHING IN THIS BOOK COMES DOWN TO A SINGLE idea and four movements, and I want to gather them here, in one place, for the days you need them close at hand.

The idea is this: experience is never neutral. It is always being authored — by intention or by neglect, but authored either way — through touchpoints, the small, deliberate, repeated points of contact through which meaning is layered into ordinary time until a place, a relationship, or a life becomes something we can actually feel.

The work is to author it on purpose, and it moves in four steps.

First, recognize. Learn to see the touchpoints already shaping your world — to walk into a room, or an inbox, or a morning, and notice that every detail is speaking, and that most of them were never decided by anyone. The simplest way in is the stranger's walk: come at the thing you care about as if for the first time, and ask what it is already saying. There is no neutral. A detail left to chance still speaks; it simply says something you did not choose.

Second, value. Understand why the small things carry so much. Attention is the whole currency, and trust is built slowly and broken quickly, because we are wired to remember the one moment that broke the pattern. The real threat is rarely disaster; it is neglect — the touchpoint nobody tended — and the surest repair is not money but noticing, and the freedom to act on what you notice.

Third, build. Make the touchpoints on purpose. Author the anticipation before anyone arrives. Make every touchpoint tell the same true story, so the experience coheres. Free the people closest to the moment to say yes. And repeat the good ones on a schedule — because a single wonderful moment is a memory, the same moment returned is a ritual, and ritual is how belonging is made.

Fourth, leverage — though the word is too cold for what it means. Protect the human, handmade, unmeasurable touchpoint as the one thing a competitor cannot copy and a machine cannot fake. Do not cut it to save money, and do not cash it in to make money. It is not the thing you trim to survive. It is the thing you were surviving for.

And beneath all four, the one thing this book is really about: presence. The capacity to be fully inside your own life, and to make another person feel truly seen. That is the rarest resource of the age we are entering, and the most valuable thing any of us will ever build — in a business, and in a life. Everything else is only the map. The room was always inside you.

# A Note on the Writing

I WROTE THIS BOOK AS A HUMAN BEING, DRAWING ON decades of experience, observation, and curiosity — and I wrote it with artificial intelligence as a thinking partner. That pairing is not a contradiction; in a way, it is the whole point. This book argues that these new tools can be a gift, if we use them to become more present and more human rather than less. Writing it was my own attempt to do exactly that: to let the technology help me organize a lifetime of material, test my thinking, and find the path through years of ideas I had carried but never quite shaped. The experiences in these pages are mine; the machine did not live them. What it did was help me carry them to you — and I've come to believe that this kind of partnership, used with intention, is something that can serve all of us.

# Acknowledgments

THERE ARE TWO KINDS OF PEOPLE IN MY LIFE.

There are those who love me and know me to the core — who accept all of it: the strengths, the weaknesses, the idiosyncrasies, and the unknowable parts every one of us carries — and who somehow still see the power and the potential of what I might yet become. And there are those who saw that potential first, though it may have had more to do with what I projected than with anything real.

This book is for each of you.

To those who love me: you know my whole history, you understand what it has cost me, and you sit in the cheering section and yell even when the team is losing. You are my faithful warriors of encouragement.

And to those I have fooled into believing I actually know what I'm doing: I love your faith in me — that beautiful, slightly blind faith that makes me believe I can really pull it off. You are the power surge when things get hard.

This book is for you.

I am a storyteller of place. And because so much of my journey was unplanned, it may have looked, at times, like I wasn't very good at reaching my destination. Not true. Life is a journey, and no matter how carefully we plan, the getting there is always full of twists and turns. So many of the places and people in this book I would never have found without exactly those detours. Thank you for sharing them with me.

To my dear friend, Janice Anderson — you are the most giving spirit I have ever known, and you remind me so much of my mother, who was adored by everyone she met. Thank you for the hours of laughter and tears and unwavering support across a lifetime. We share so many of the threads that have woven our lives together, and the tapestry we've made of all that time is something I will cherish always.

To Donna Ashcraft, my friend and my mentor — thank you for showing me what a powerful woman can be, and for doing it with such kindness and sophistication. You are a masterclass in professionalism with a Southern twist: sweet tea with a dash of spice.

To Peter Pappas, President of Pappas Properties, Philip Tate of Luquire George Andrews Agency, and Kate Caltis Asst Vice President at Pappas, from my Birkdale Village, Berewick, and Waterdance years — your belief in my ideas and your shared passion for placemaking made that work a joy, and shaped so much of what made those projects succeed. And to Patrick Schneider, photographer extraordinaire, who captured the heart and soul of Birkdale Village in images that will hold those memories for generations.

To Bob Nelson, whose leadership as General Manager at The Wharf, made my job so much more joyful. Your humor and dedication made every day fun. And to Beason Wilkes, Director of Development for this $800M project — your memory remains as another beacon of encouragement to pursue a task that I would otherwise never have taken on.

To my family — this book is my way of explaining what I've been doing all these decades, and why it has mattered so much to me. You knew the facts; I hope now you also feel the feelings, some of which I only made peace with in the writing. I hope that each of you, at some point in your own life, takes the time to rummage through your own chapters — to see what quietly reveals itself once the stories begin to come together.

And to you, the reader, thank you for taking a chance on me. I hope this book helps you live a little more present, and a little more on purpose, in your own life. Notice the little stuff. Savor the details. And then, give yourself some grace. Nobody gets through this life without a little scar tissue — I only hope yours came from embracing every touchpoint.

# About the Author

Elizabeth Lord is a brand strategist and storyteller of place whose career has been devoted to shaping the way people experience where they live, gather, and return.

Her work spans more than two decades and includes signature projects such as Birkdale Village, the Nantucket-style town center outside Charlotte, North Carolina, and The Wharf in Orange Beach, Alabama. She has also provided corporate marketing leadership for major real estate and development brands across the Southeast and for a luxury international real estate brokerage.

After nearly fifty moves, she came to understand that belonging is not a slogan but a feeling to be built with intention — sometimes on ground that once held nothing but a cow pasture. That insight has shaped a career centered on one question: how a place, a brand, or a moment can be made to feel like home. She lives in Greenville, South Carolina. Every Touchpoint is her first book.

*See the places behind these stories at www.langworthylord.com.*